BIOLOGIC VARIATION in HEALTH and ILLNESS

BIOLOGIC VARIATION in HEALTH and ILLNESS

race, age, and sex differences

Theresa Overfield, RN, MPH, PhD

Professor and Director of Research
College of Nursing, Brigham Young University
Research Associate Professor
Department of Anthropology, University of Utah

ADDISON-WESLEY PUBLISHING COMPANY
Nursing Division • Menlo Park, California
Reading, Massachusetts • Don Mills, Ontario • Wokingham, UK
Amsterdam • Sydney • Singapore • Tokyo • Mexico City
Bogota • Santiago • San Juan

Sponsoring editor: Nancy Evans
Production and copyeditor: Zipporah W. Collins
Cover and interior designer: Alice Klein
Proofreader: Linda Rageh

Library of Congress Cataloging in Publication Data

Overfield, Theresa, 1935–
Biologic variation in health and illness.

Includes index.
1. Human biology. 2. Variation (Biology)
3. Race. 4. Medical anthropology. I. Title.
[DNLM: 1. Aging. 2. Racial Stocks. 3. Sex Characteristics. 4. Variation (Genetics) QH 401 096b]
QP34.5.094 1985 611 85–1338
ISBN 0–201–12810–1

ABCDEFGHIJ–AL–898765

Addison-Wesley Publishing Company
Nursing Division
2725 Sand Hill Road
Menlo Park, California 94025

Contents

Foreword

Once upon a time there was a country where everyone had always been Black and no one knew that people could be any other color. One day a small boatload of White people was shipwrecked and cast up on the shore. The physicians of that land immediately began to argue among themselves about whether White skin was a congenital defect or the result of some kind of disease. A horrifying thought occurred to one of them: if it was a disease, it might be contagious. The unfortunate castaways were immediately put in strict quarantine. But eventually the puzzled physicians set about trying to find a cure for this strange and unsightly condition. They did not succeed.

None of us in this modern world of refugees and travelers would be as innocent of skin color differences as the well-meaning physicians of this legend. However, there are many other differences between populations derived from different parts of the world. Some are plain to see, like differences in the shape of the facial features, though we often pay little attention to these if the skin color gives us a "name" to apply to someone. Other traits differ in ways that we recognize only by laboratory tests; as Dr. Overfield points out, some of these differences may mislead us in diagnosis or confuse us in treatment. The whole human species, of course, is tremendously variable. Even within one nation, no matter how isolated, even within one family, we find innumerable differences between individuals. In ways that we do not fully understand, these differences have become partially sorted out according to geographic area (or, as we must say in the modern world, area of ancestry). In the case of body size, we find different averages in different populations of the world, with every population having much variation in individual size as well. Two populations may have the same blood groups present, but in different percentages. Some of the differences correlated with area of ancestry probably arose many thousand years ago, when small bands, perhaps a few families, left a group and went out to found new tribes. Their individual and family characteristics became the heritage of what later became large populations. Thus population and racial differences are, in a sense, the lengthened shadow of individual differences.

Even individual differences within a reasonably uniform population may be disconcerting to the medical profession. What is "normal"? How much variation from the "normal" can we ignore before we decide that "something has to be done"? The skeletons of prehistoric American Indians, who led a much more active and rigorous life than most of us, often show a twisting of the upper end of the femur that a modern orthopedist considers abnormal. In this case a racial difference tells us that we are too ready to consider what is *unusual*, in the populations we are used to dealing with, as *abnormal*. It is not surprising that the norms set by Western medical traditions were based on the people of the countries in which those traditions developed. But even within these areas, norms may have been derived from special segments of the population. For instance, it was supposed for many years that hemoglobin levels ought to be the same for the two sexes, and many women were classed as "anemic" although their lower hemoglobin levels were quite natural and not due to any dietary deficiency or pathologic condition. For this reason, Dr. Overfield discusses, along with racial differences, individual differences and differences by sex and age. These are inseparable parts of the great tapestry of human variation.

Many people are reluctant today to admit any but the most obvious and visual of racial differences. This embarrassment stems partly from ignorance and partly from a floating anxiety about social inequities. Obviously, if we do not admit that differences exist, we have a good argument against those who claim (a) that they are different from us, (b) that they differ by being better than we are (according to their own standards, of course), and (c) that because of the differences they deserve special privileges. But in the end, the reality of differences, aside from any value judgments about it, cannot be concealed. We can only demand that its existence not be misinterpreted or misused. Sooner or later we must fall back on the real argument, the moral argument, which is that every human being is entitled to all the opportunities he or she can take advantage of and is entitled to the best possible help when needed. Dr. Overfield discusses human individual and group differences so that we can know how to give the best help to everybody.

Alice M. Brues
Department of Anthropology
University of Colorado
Boulder, Colorado

Preface

This book was conceived during a course I have been teaching at the University of Utah since 1976, called "Biological Variation in Health and Illness." The need to offer such a course first became apparent to me after I took the course, "Human Races," taught by Dr. Alice Brues of the Department of Anthropology, University of Colorado. Although her course did not relate specifically to health and illness, my training and experience with Eskimo, Indian, Black, and Mexican-American clients had alerted me to the health implications of what I was learning. I determined early on to make this information available to health care providers.

The Importance of Biologic Variation in Health Care

Familiarity with external and internal biologic variations that normally occur enables heath professionals to make the transition from discomfort to an informed awareness in the presence of those who differ in appearance. Health workers can get beyond the "all-Chinese look-alike" feeling and intellectually satisfy their desire to stare by understanding why people differ in their external characteristics. With time and familiarity, discomfort usually lessens, but information on biologic variation will speed up the process considerably. It is preferable for health personnel to concentrate their energies on individuals and their health problems rather than on their own problems of culture shock and discomfort.

An understanding of how and why groups vary biologically can alleviate ethnocentrism and cultural shock to some extent. Ethnocentrism is due partially to a feeling of biologic superiority. When biologic superiority is defined as a more successful adjustment to the environment in which a person lives and when it is measured by the ability to produce children to carry on the person's genes, health care providers may recognize that superiority in the very people they were feeling superior to.

The Audience for This Book

This book is designed for graduate students in medical anthropology, medicine, and nursing. There are no prerequisites to an understanding of the book. Readers lacking a background in population genetics, for example, will find a survey of the essential facts and theory in the first chapter.

Currently, research findings about biologic variation as it affects health and illness are scattered among several journals not regularly read by nurses and physicians. Most of the research on biologic variation has been done by physical anthropologists and is published in their journals. This book is a compilation of that research as it pertains to health and illness.

In the health field, the implications of racial variation have been overshadowed by newfound interest in cultural differences. It is my contention that human variability must be seen in more than one dimension. Knowledge of both cultural and biologic variation is an essential component in providing adequate health care to individuals of differing cultures or races.

Organization and Approach

This book summarizes what is known about biologic variation as it relates to health and illness. Many misconceptions about racial uniformity—and diversity—will be exposed, especially those that may affect health care. The chapters are arranged to reflect my way of thinking about biologic variation. Chapter 1 presents an overview, from an evolutionary perspective, of biologic variation and its sources. Chapter 2 contains a discussion of the most obvious racial differences. These variations are used to label and categorize individuals, so it seems important to start with them and to show that race is more than skin deep; there is more to it than meets the eye.

Chapter 3 examines biologic differences throughout the life span. Chapter 4 discusses influences of the internal environment (biochemistry and disease susceptibility), while Chapter 5 looks at present-day influences of the external environment on biologic variation.

Nervous system variation is discussed in a separate chapter, Chapter 6, because of misconceptions about biologically determined behavior. Another area full of misconceptions, sexual variation, is the topic of Chapter 7, which shows that, excluding obvious anatomic and hormonal differences, there are more variations between the sexes than among the races.

Chapter 8 is a detailed look at the Sioux Indians from the perspective of biologic variation. The chapter illustrates the type of description needed by health personnel when they work with other racial groups. The author of chapter 8, Barbara Criss, a nurse-midwife, has worked with the Sioux for the past 10 years and is currently completing a doctorate in transcultural nursing at the University of Utah.

The Afterword notes the paucity of good research on biologic variation in general and the lack of studies of some groups and certain topics. It challenges the health community to become involved in research in this area.

Theresa Overfield

Acknowledgments

This book had a seven-year gestation period; consequently many people helped and influenced me during the writing of it. First I wish to acknowledge Professor Alice Brues, Department of Anthropology, University of Colorado, who introduced me to racial variation and who started me on the path that led to this book. Next I wish to thank Dean Elaine Dyer, College of Nursing, Brigham Young University, who has supported my endeavor during the past few years. Several nursing colleagues read and critiqued chapters as they were written: Mary Duffy, Ruth Kennington, Marilyn Lyons, and Margaret Armstrong. John McCullough, Chairman, Department of Anthropology, University of Utah, read and critiqued early chapters and encouraged me along the way. Several graduate students helped me search the literature: Richard Boortz-Marx, Wilma Buerger, and Judy Brown from the Brigham Young University College of Nursing, and Karen Rice from the University of Utah Anthropology Department. The final draft of the manuscript was reviewed by Barbara Reynolds Caldwell, Villanova University; Betty L. Chang, University of California, Los Angeles; Mattie Collins, Borough of Manhattan Community College; Audrey Koertvelvessy, Ohio University, Athens; Frances Ouellette, Boston College; Martha Primeaux, University of Oklahoma; and Mary Opal Wolanin, Geriatric Nursing Consultant, Tucson. Their comments were constructive and helpful. Nancy Evans, sponsoring editor at Addison-Wesley, recognized the need for this book and had faith in its author. Zipporah Collins, production and copyeditor, got me through the final frantic weeks of readying the book for the press. She made the experience almost pleasant. Special thanks are due to Barbara Criss, graduate student in nursing, University of Utah, who read and commented on the manuscript and helped keep me practical. Evalyn Bowerman deciphered my writing and typed the first draft of my book. She did so with dispatch and sent along encouraging notes. Julie Chadwick Collins typed the last drafts on the word processor and kept a cheerful mien despite the many revisions I gave her. Don Norton of the Brigham Young University English Department, and one of his students, Karl Hayes, read the manuscript and gave me editorial advice. Finally I wish to thank Ruth Hensen who taught me how to write and who read and edited the penultimate draft.

chapter 1

Biologic variation introduced

Little-known physiologic differences among groups produce different responses to some healthy and diseased states. These different responses are usually overlooked by both those giving health care and those conducting research on normal body responses to health and illness.

This oversight occurs partially because biologic variation is largely the concern of physical anthropologists, whose journals are little read by health professionals. Also, anthropologists have studied race for their own reasons, not realizing the relevance of their findings to health care.

Research into biologic variation has focused on differences between racial groups. It grew out of the interest of physical anthropologists in race, a concept whose existence or nonexistence continues to be debated by the anthropologists. The common current view is that use of the term *race* is legitimate if applied to a geographic population or to an isolated smaller group.

Health care workers can no longer ignore the body of research on the biology of race. The political and social upheavals of the 1960s and 1970s have prompted a number of legislative regulations that require equal treatment of the various groups in the United States. The emphasis has been on what races have in common. While this is commendable, it has also helped obscure the real differences that exist among races. Equal in rights is not the same as equal in treatment. Health care must take into account known biological differences among people of different races.

Human beings are similar: they are of the same species, and they enjoy the same privileges under law. But they differ in many important ways: in growth and development rates, in enzyme systems, in disease susceptibility, and probably in nutritional requirements. Body proportions differ among the races, and there are even anatomic differences among some groups. People are not all alike under the skin.

Racial differences have developed as a result of earlier responses to past environmental influences. Present-day differences can be viewed as a map of the evolutionary past. Understanding how each group evolved in relation to its environment makes it clear that facial features, skin color, body size and shape, even enzymes are the results of biologic adjustments that each group's ancestors were able to make to their environment. For example, it has been postulated that the original skin color of human beings was black; white skin resulted from mutations and environmental pressures incurred by those living in cold, cloudy, northern Europe (Neer, 1975). Light skin is better able to synthesize vitamin D on infrequent sunny days; conversely, black skin is a positive trait where protection from the tropical sun is a factor but is a neutral to negative trait in other circumstances.

The majority of nurses and physicians in the United States today, members of the White race, have unconsciously incorporated their own ideas of what is "normal" into their management of patients. Most studies of biologic baselines for growth and development, nutrition, and laboratory values have been carried out on Whites, producing standardized norms that may not apply to other racial groups. White values are accepted as normal; other values are considered abnormal—rather than simply non-White. Hemoglobin values are an example: healthy Blacks and Eskimos usually have hemoglobin values approximately 1 g below those of Whites (Maynard, Hammes, 1970; Garn, Smith, Clark, 1975). Frequently these groups are treated unnecessarily with iron supplements.

If health care practices do not reflect the differences among racial groups, inadequate and possibly harmful treatment may result. For instance, while a small amount of iron supplementation may not hurt, advocating milk intake for individuals deficient in the lactose enzyme may cause real harm. Approximately 80–90% of Blacks, Native Americans, and Orientals are lactose enzyme deficient; for those people milk consumption can cause diarrhea, abdominal distention, and excessive gas production. Even the notion that women need large amounts of calcium during pregnancy and lactation is something I seriously question.

The material in this book is presented from an evolutionary perspective; the development of the great variety seen in human beings is a form of evolution. Consequently, an understanding of race and racial variation, including a review of human genetics and a nonmathematical survey of related concepts in population genetics, is necessary.

RACE AND RACIAL VARIATION

Human variation can be viewed at various levels: race, the largest category; racial subgroups, an intermediate category; and individual, the smallest category. This book deals with the first two categories but not with individual variations, and it describes only fairly common variations. Although individual variations have significance for health, these are more suitable for a medical genetics text. Rare conditions make up a good part of medical genetics. This book looks at health-related variations that affect large numbers of individuals.

Succinctly put, *race* is "a division of a species which differs from other divisions by the frequency with which certain hereditary traits appear among its members" (Brues, 1977, p 1). In another dimension, "race is a breeding population . . . which for reasons of geography or culture mates largely within itself" (Goldsby, 1977, p 4). Of course the fact that it is a breeding population accounts for the similarity of its members when contrasted with other breeding populations. There is general agreement among students of race that the characteristics that define a specific race will not define any specific individual from that race. For example, Yupik Eskimos can be described as having a 75% gene frequency of M in the MN system; no Rh-negative blood; a frequency of 63% for dry earwax; and in the ABO blood group 61% are O, 23% A, and 16% B. The most common genotype for a Yupik would be blood types O, M, and Rh-positive and dry earwax. While some Yupik Eskimos would actually have this array of traits, even more would differ.

Another important qualification is that very few races, or racial subgroups, mate completely within their own group. There are really no true boundaries between races. *Gene flow* is the term used to describe the lack of real boundaries and the concomitant sharing of genes through mating that occurs along these imprecise boundaries. In the past, when populations were more stable, gene flow took place because of marriage customs or because of war-induced rape or slavery. Since modern times, boundaries have become less distinct because of increased mobility and decreased cultural constraints.

In this book, I generally recognize three major groups: White, Black, and Oriental/American Indian, as does Brues (1977). American Indians and Orientals are considered to belong to the same major group. Variations in the intermediate-level groups, or racial subgroups, often have a geographic or nationality label, whichever fits better—for example, northern people, Mediterranean people, the Irish, and Sioux Indians.

Readers need to keep in mind that a specific variation may or may not be present in any one individual. This book merely points out the possibility that an individual's ancestors were generally known to have the characteristic.

The terms *biologic variation* and *racial variation* are used interchangeably here. Although the word *race* is somewhat unfashionable today, it is a good short word and a legitimate term when properly defined.

Races are not static; they have evolved in response to earlier environmental influences, and they continue to evolve. Local groups have responded to the challenge of their environment in many different ways. However, similar environments did not necessarily produce the same response in different local groups. For instance, northern Europeans differ from northern Chinese in numerous ways. Their climates are similar, but their evolutionary histories have differed.

HUMAN GENETICS—A REVIEW

Humans have 46 chromosomes in 23 pairs: 22 pairs are called *autosomes*; one pair is the *sex chromosomes*, X and Y (XX for female, XY for male). The only known function of the Y chromosome is to produce a male. Traits are *sex-linked* not because of the Y chromosome but because they are exclusively on the X chromosome; in fact, they are said to be *X-linked*. An example of this is hemophilia, a disease that occurs mostly in males. Males have only one chance to receive the normal gene, which is located on the X chromosome. Females have the disease rarely, because they have two chances to inherit a normal gene, since they have two X chromosomes. Some traits are *sex-limited*; these traits are expressed only in the presence of the specific sex hormone, though the genes involved are not located on the X chromosome. Male pattern baldness is an example. It does not develop unless testosterone is present in addition to the genes for baldness.

Chromosomes are made up of genes, which are in turn composed of the major chemical of heredity, deoxyribonucleic acid (DNA). Genes are chemical substances with specific functions; they are the templates that produce the enzymes that run the body processes.

The specific placement of genes on chromosomes is not yet well documented for humans, mainly because of the millions of genes involved. For instance, the X chromosome alone contains at least 93 genes, and the locations of most of these are only tentatively known.

At any one locus on a chromosome there may be one of two or more genes called *alleles*. For example, A, B, and O are the alleles for the ABO blood type locus. Because an individual receives two genes (one on each chromosome of the pair of chromosomes), they may be the same or different. When the genes are the same, the individual is called a *homozygote*; when the genes are different, a *heterozygote*.

The action of genes, individually or in concert, produces an effect called the *phenotype*. Phenotypes may represent the action of one set of genes or many. Phenotype is distinguished from *genotype*, which is the specific pair of alleles that makes up a trait. Genotype is the more specific term. Phenotypes may come about in a variety of ways; genotypes are specific to a single set of alleles. The ABO blood system illustrates these terms: the four blood types are A, B, O, and AB. In actuality, there are six different blood types: AA, AO, BB, BO, OO, and AB. No one has yet developed a laboratory test to distinguish between AA and AO, or between BB and BO, because A and B are dominant genes. Consequently, we have A and B phenotypes, O or OO and AB genotypes. Earwax is another trait that may be used as an example. There are two types of earwax in humans, wet and dry. The allele for dry earwax is recessive; therefore, we know the genotype of all persons with dry wax (ww). Wet wax is dominant, so all homozygote wet wax persons (WW) and all heterozygote wet wax persons (Ww) appear the same; hence, we can only call these phenotypes. *Phenotype* is also used to designate an effect produced by several genes working together. For instance, though eye color is produced by the interaction of several genes, we may call blue eyes a phenotype.

Manifestation, that is, the expression of a gene, may be dominant, recessive, or somewhere in between. *Dominance* means that when one allele is paired with a different allele, it overshadows the other. *Recessive* means that the effect of the gene is manifested only in the homozygote state—when both alleles are the same. Some genes produce a blending effect in the heterozygote state; an intermediate phenotype results. It is thought that many, if not most, genes fit into this category. At present researchers do not have the technical sophistication to distinguish the subtle effects of the so-called recessive genes.

Most traits are *polygenic*, that is, they are produced by the cumulative effects of genes at many loci. Examples of these are height, skin color, hair color, and tooth structure. Because most traits are polygenic, what people generally observe are phenotypes; they cannot determine the genotypes or even the number of loci involved in most traits.

An additional factor is *pleiotropy*, the capacity of a gene to influence a variety of phenotypic traits. Skin, hair, and eye color, tooth structure, and many more traits involve pleiotropy. An example of pleiotropy is seen in the condition phenylketonuria, where the abnormality produces both impairment of the nervous system and lightened skin, hair, and eye color.

POPULATION GENETICS—AN OVERVIEW

The variations in human populations set forth in the following chapters occur for many reasons. Researchers do not understand all the reasons for the occurrence of specific traits, but they know in general why these differences arise. The biologic variations in present-day groups are the result of natural selection, and of other processes of genetic change, that took place in the past. These variations provide the raw materials for selection in the present, thus giving the species genetic flexibility and the possibility of a future.

In order for nature to choose fit individuals to pass on their genes to the next generation, there must be a variety of individuals. Natural selection rarely involves a dramatic choice of individuals for survival. People commonly wonder how this or that individual has managed to survive, let alone produce offspring. Sometimes, in fact, the more inept individuals seem to have the largest families. These thoughts illustrate a common misconception about natural selection. It is not "survival of the fittest," as Darwin is often quoted, but survival of *the fit*. Even people's judgment about and exasperation over the fruitfulness of an "obviously inferior type" is an enormous presumption. Who are they, themselves potential candidates for selection or elimination, to judge which genotype will do best under current and future conditions? Their own sensitivity about being crowded by such a person may in itself be a nonadaptive trait, one that could ultimately result in their doom.

Many species may have became extinct because they could not maintain enough variability for natural selection to work. The saber-toothed tigers did not make it, the great whales are becoming extinct today, and several human races have died out or been absorbed into the larger gene pool. *Homo sapiens neanderthalensis* disappeared a few thousand years ago; the Tasmanian race off the coast of Australia died out shortly after White colonization. Several Indian tribes on the North American continent did not survive long after Europeans arrived. These groups failed because one or more of the conditions necessary for survival were exhausted. These "conditions for survival" will be defined shortly.

The inherited variability from which selection chooses is derived from several sources. The ultimate source of all variation is mutation, but other mechanisms—genetic drift and gene flow—add to it. *Mutation*, a random, inheritable change in the genetic material (DNA), is rare. It can be produced by specific mutagenic agents: x-rays, certain drugs, and certain chemicals. It also may occur for no known reason. By liberal estimates, normal mutation rates at each locus probably do not exceed 1 in every 5,000 generations (Brues, 1977). Most mutations are deleterious, some are neutral, and only an occasional one is advantageous. The structures of our genes and chromosomes are so complex and have evolved over so many generations that the random change produced by a single mutation has only a very small chance of improving the structure. Nevertheless, mutation is the ultimate source of variability.

Genetic drift is another mechanism that makes populations different. Genetic drift occurs because of random fluctuations in gene frequencies in small populations. It is akin to the error that arises in using small sample sizes of any kind. A small sample is rarely exactly representative of the population from which it is drawn. Small populations experience random fluctuations in gene frequencies, as large populations do, but a gene at low frequency in a small population may disappear altogether because of genetic drift. In a large population such a gene is better protected from random extinction.

Genetic drift also occurs when a small group splits off from a larger population and proceeds to mate within itself. The genes of the group's founders will be passed on. After several generations, the new small population will probably differ from the larger population from which it split off, because the founders were not a representative sample and were such a small group that random fluctuations accentuated the original difference.

When people move about and settle in places different from where they were born, they move their genes along, too. This changes the gene frequencies in populations by introducing more variability. The American Black population differs from the African Black population because of gene flow and genetic drift. There is about a 20% admixture of White genes in American Blacks (Reed, 1969). Moreover, because Blacks who were brought here by slavers were not a representative sample of all African Blacks, genetic drift has compounded the differences.

Population gene frequencies remain stable over generations, provided that no mutation, gene flow, genetic drift, or selection occurs. This principle is called the Hardy-Weinberg equilibrium law, the basic law of population genetics. It explains mathematically why gene frequencies tend to remain the same. The variations mentioned in this book are caused by departures from Hardy-Weinberg equilibrium. If it were not for these departures, sometime in humanity's racial past, humans either would be one large, homogeneous race worldwide or would be extinct. The latter possibility is the greater. Thus, variability helps ensure survival of the species.

Because variability, the topic of the remaining chapters in this book, is essential for natural selection, it is necessary to know a little more about selection. *Selection* is positive differential reproduction, such that the reproductive capacity of one individual or group is higher than that of another. The only valid measure of fitness is the number of offspring produced who go on to produce the next generation.

Even though the ultimate measure of selection is reproductive capacity, selection can happen at any time in the life cycle. Individuals can be eliminated before gestation is complete, after birth and before reproducing, or during the reproductive years; in each case they produce fewer (or no) offspring. Death after the reproductive years is not directly a part of natural selection, but the death of a parent before children are able to care for themselves may cause fewer offspring to survive. Survival of adults until their offspring are self-reliant, especially under adverse conditions, probably accounts for the presence of some genes for longevity. Really long life, into the 80s and 90s, is not explainable by natural selection.

Two major types of selection produce effects on the frequency of genes in a population. These are directional and stabilizing selection. Using the simplest case of a two-allele system, Aa, there are three possible genotypes: AA, Aa, and aa. If one of the homozygotes (AA or aa) is the most fit genotype, for example, AA, then gradually all of the a genes will be lost. This is *directional selection*, the kind that comes to mind in the phrase "survival of the fit." The relative superiority of one genotype over the others determines how long the process will take. If selection is due to some new condition or disease, where only homozygote AA individuals survive, then selection will eliminate all a genes in one generation. Directional selection is usually more subtle than that. Using the example of sickle cell anemia, homozygotes for the sickling gene usually die before reproducing. Heterozygotes are at a slight disadvantage in a malaria-free environment, especially at high altitudes or under conditions of physical stress. Consequently, heterozygotes will be less fit than normal hemoglobin homozygotes. Because the disadvantage is only slight in a malaria-free environment, elimination of the sickling gene takes many generations. American Blacks have a lower sickle cell gene frequency than African Blacks, even taking into consideration the admixture of White genes. This difference in gene frequency is very probably caused by directional selection (Bodmer, Cavalli-Sforza, 1976).

Stabilizing selection comes about when the heterozygote is at an advantage; consequently the heterozygote produces more offspring than either homozygote. Again the classic example is the sickle cell trait, but in a malarial area. The homozygote for sickle cell dies before producing children; the homozygote for normal hemoglobin has more severe malaria than does the heterozygote. It is thought that, as the malaria parasite attempts to reproduce in an individual who is heterozygous for the sickle cell gene, the infected red cells with the sickle cell defect are destroyed before the malaria parasite can complete reproduction. This reduces the parasite load. These red cells sickle when they are parasitized and are then eliminated from circulation by phagocytosis (Roth et al, 1978).

Stabilizing selection keeps two alleles present in a population at unvarying frequencies. It is interesting to note that the best-documented type of selection is the one in which little or no change in gene frequencies occurs. Directional selection, because it is usually very subtle or very transient, is difficult to document in living peoples. Investigators may eventually see verifiable examples of directional selection, but, unless they are present now and are hidden by their subtlety, people will not be happy to discover them. An easy to imagine and very possible example would be when (not if) some microbe evolves that causes a worldwide epidemic, killing a good portion of the human population; the survivors would be those selected. I hope a population geneticist is among the survivors, to document the directional selection!

When both of the homozygotes and the heterozygote have the same survival rates, no selection occurs, and gene frequencies remain the same. This is the condition assumed in the Hardy-Weinberg equilibrium law.

As mentioned earlier, groups become extinct when the conditions necessary for survival are not present. The same applies to genes. Extinction of a gene results from directional selection. Extinction, especially of genes, can also result from genetic drift. Both these conditions are usually compounded by inbreeding, which further decreases variability. Safety from extinction is not assured even in the presence of great variability, however. The environment may change in such a way that the organism has no buffer against it: a severe climatic change, a new plague, or a nuclear holocaust. Rather than end on a depressing note, though, let me say that humanity has made it so far—perhaps this luck will continue!

REFERENCES

Bodmer WF, Cavalli-Sforza LL: *Genetics, Evolution and Man.* Freeman, 1976.

Brues AM: *People and Races.* Macmillan, 1977.

Garn SM, Smith NJ, Clark DC: The magnitude and the implications of apparent race differences in hemoglobin values. *Am J Clin Nutr* 1975; 28:563–568.

Goldsby RA: *Race and Races*, 2nd ed. Macmillan, 1977.

Maynard JE, Hammes LM: A study of growth, morbidity and mortality among Eskimo infants of western Alaska. *Bull WHO* 1970; 42:613–622.

Neer RM: The evolutionary significance of vitamin D, skin pigment and ultraviolet light. *Am J Phys Anthropol* 1975; 43(3):409–416.

Reed TE: Caucasian genes in American Negroes. *Science* 1969; 165:762–768.

Roth EF et al: Sickling rates of human AS red cells infected in vitro with *Plasmodium falciparum* malaria. *Science* 1978; 202:650–652.

chapter 2

Surface variations and anatomic differences

The racial differences most often noticed are skin color, facial features, hair type, and body proportions. This chapter discusses many of these features, plus less obvious surface and anatomic characteristics. Many more surface and anatomic characteristics contribute to the description of racial variation than this chapter covers; it includes primarily the characteristics pertinent to health care. Little-known but common normal variations within races, other than the White race, are also included to help distinguish normal variation from pathology. For details on racial differences with no known health impact, see the text *People and Races* (Brues, 1977).

SURFACE VARIATION

Racial variation comes packaged in skin. The color of the skin is one of the major ways people categorize each other; other surface characteristics subtly add to this categorization.

Color

Skin Skin color is the most obvious way in which populations vary worldwide. Eye and hair color are much less variable; most humans are dark-haired and dark-eyed. The people of northern Europe are the only ones who vary much in hair and eye color.

The source of skin color variability is the pigment produced by the melanocytes, which are located in the epidermal layer of the skin. The melanocytes originate in the neural crest near the embryonic central nervous system and migrate into the fetal epidermis (Wasserman, 1974). This origin explains two things about pigment. First, mongolian spots are migratory leftovers that lingered in the lumbar-sacral region at a greater than ordinary depth, causing the blue effect. Mongolian spots are present in 90% of Blacks, 80% of Orientals and Native Americans, and 9% of Whites (Jacobs, Walton, 1976). They occur on the buttocks, back, and, occasionally, abdomen, thighs, and arms. Health professionals unfamiliar with these blue spots could mistaken them for bruises and could misdiagnose child abuse.

Second, the neural origin of the melanocytes explains their shape. They resemble neurons with dendritic processes that insert into epidermal cells. These processes inject melanosome granules, which contain the melanin pigment, into the epidermal cells (Szabo, 1975).

It is common knowledge that dark-skinned individuals are protected from the effects of sunlight to a greater degree than light-skinned individuals. A reason for this may be the dispersal of the melanosomes from the basal cells up to the stratum granulosum. In Whites, Orientals, and American Indians, melanosomes are incorporated singly (Szabo et al, 1969). The melanosomes of Blacks are longer and broader than those of other races. It is possible that the even dispersal of the melanosomes gives the skin a more uniform and dense color, which provides better protection from the sun. Blacks do become sunburned, but only after more exposure than for Whites. Dark-skinned individuals also have a lower incidence of skin cancer than light-skinned individuals.

Although melanin provides some protection against cancer and sunburn, dark-skinned individuals have other problems that are related to skin pigment. Melasma, the "mask of pregnancy," is more prevalent in dark-skinned individuals. The exact cause of melasma is unknown, but the condition occurs during pregnancy or during use of birth control pills. The forehead, malar prominences, and upper lip become hyperpigmented. If it appears during pregnancy, melasma usually fades 3 to 6 months postpartum. If it is triggered by the use of birth control pills, it may never fade (Resnik, 1967). Women should be told about this side effect. If hyperpigmentation becomes noticeable, health care providers might suggest that a woman discontinue the pills and substitute another method of birth control. Use of strong sunscreening lotion might also help.

Other areas of the skin affected by hormones and, in some cases, differing by race are the sexual skin areas: the nipples, areola, scrotum, and labia majora. In general, these areas are darker than other parts of the skin in both adults and children, especially in the Black and Oriental races. Racial differences in the areola have been demonstrated in a group of White, Black, Native American, and Chinese women (Pawson, Petrakis, 1975). As expected, Black women have the darkest areolas; Native Americans and Chinese are intermediate; Whites have the lightest. A more surprising finding is that the contrast between areolar pigmentation and the remainder of the breast is greatest for Native American women. Black and Chinese have intermediate degrees of contrast, with White women showing the least contrast.

Contrary to popular clinical belief, there is no apparent correlation between lightness of nipple pigmentation and nipple damage during breastfeeding. Light-skinned women do not have more nipple soreness or cracking than do dark-skinned women (Gans, 1958; Brown, Hurlock, 1975).

Vitiligo is a common pigment problem. It is thought to be an autosomal, dominant condition in which the melanocytes become nonfunctional in some areas of the skin, resulting in unpigmented skin patches. It affects over 2 million persons in the United States; in some countries, as much as 5% of the population is affected. There appears to be no sex or race predilection, but vitiligo poses a problem mainly for dark-skinned individuals. If light-skinned individuals lose pigment, avoidance of suntanning will help by decreasing the contrast between pigmented and unpigmented skin. Individuals with vitiligo also have a higher than normal chance of developing pernicious anemia, diabetes, and hyperthyrodism. These factors probably indicate some deeper genetic problem (Bor, Feiwel, Chanarin, 1969).

Another pigment condition showing racial variation is leukoedema, a grayish white benign lesion occurring on the oral buccal mucosa. It is present in 68–90% of Blacks and 43% of Whites, but it is rarely seen in the people of India (Martin, Crump, 1972). The condition increases with age in all races.

Oral hyperpigmentation also shows variation by race (Wasserman, 1974). Absent at birth, it increases with age. By age 50, 10% of Whites and 50–90% of Blacks will show oral hyperpigmentation. It is thought to be caused by a lifetime accumulation of postinflammatory changes in the mouth.

Eyes There is variability in the color of the iris and in retinal pigmentation, with a positive correlation between the two areas: darker irises have darker retinas behind them (Hoffman, 1975). Retinas vary from a light red-orange to dark brown or gray. Individuals with light retinas generally have better night vision but can suffer pain in an environment that is too light. The illumination of rooms needs to be accommodated to individuals who are more light-sensitive than others. They may want rooms kept dark during the day, and be able to wander around at night with little illumination. The opposite conditions could be needed for individuals with dark retinas.

Albinism and phenylketonuria There are two medical conditions involving pigment changes that vary by race: albinism and phenylketonuria. Albinos have melanocytes, but their melanocytes do not produce melanin pigment. Albinism is a rare condition in most races. The prevalence rate is between 1 per 10,000 and 1 per 60,000 for most groups; a few American Indian tribes constitute exceptions. The Hopi, Zuni, and Jemez, all southwestern American Indians, and the Cuna, a Central American tribe, have a remarkably high rate of albinism, 1 per 200 (Woolf, Dukepoo, 1969).

Phenylketonuria, another autosomal recessive condition, is most common in northern Europeans and is rare in Blacks. The frequency in the United States is 1 per 20,000, but it ranges from a high of 1 per 4,000 in Ireland and Scotland to a low of 1 per 29,000 in Sweden (Goodman, 1979). This condition is caused by a metabolic defect that allows phenylalanine to accumulate rather than being converted to tyrosine as it should, thus causing toxic effects to the brain and central nervous system. Tyrosine, the amino acid involved in albinism, is also involved here; because there is less tyrosine, skin color is lighter, and afflicted individuals have blue or other light-colored eyes, whatever their race.

Secretions

Sweat and sebum The apocrine and eccrine sweat glands and the sebaceous glands vary by race and are significant for health. Most sebaceous glands open into hair follicles; a few open directly on the surface. Although there are no racial differences in the composition of the sebaceous secretion, there is evidence that Blacks have 60–70% more of it on their hair than do Whites (Nicolaides, Rothman, 1953). The oily material excreted onto the hair fiber as it grows is significantly thicker in Blacks than in Whites. This fact should be considered by health care workers who make judgments about or teach individuals about personal hygiene.

The eccrine sweat glands, which play an important role in thermoregulation and fluid balance, are distributed over the entire body and open directly onto the skin surface through pores. There are 2–3 million of them per individual. The eccrine glands on the plantar and palmar surfaces are under the control of the nervous system; the rest respond to thermal and fluid needs. There is little variation in the number but much variation in the activity of the sweat glands. This variation is due to both environmental and individual adaptations, not to race. The Japanese born in Japan who migrate to the tropics have the same number of functioning glands as those who live their entire lives in Japan. But the Japanese born and raised in the tropics have significantly more functioning sweat glands than either of the groups born in Japan (Kuno, 1956). Similar evidence is available on Solomon Islanders, compared with Whites and Japanese (Garcia et al, 1977).

Eskimos have made a unique eccrine gland adaptation to their environment. They sweat less than Whites on their trunks and extremities but more than Whites on their faces (Schaefer et al, 1974). This strategy allows thermoregulation without much dampening of the clothes, which would decrease their insulative capacity and pose a serious threat to survival.

The amount of chloride excreted by the sweat glands differs according to race. Blacks have lower salt concentrations in their sweat than Whites do (Batson, Young, Shepard, 1962). Whites who are acclimatized to the tropics show a lower concentration of sweat chlorides than nonacclimatized Whites. The climate can be a confounding factor. A study of Ashkenazic and Sephardic Jews demonstrates that the Ashkenazim, of European heritage, have a lower percentage of sweat chlorides than the Sephardim, of North African and Middle Eastern heritage (Levin et al, 1966). These findings indicate that there is both a genetic and an environmental influence on the level of sweat chlorides. See chapter 5 for more details on environmental variation.

A result of the differences in salt concentrations in sweat is a significant difference in galvanic skin resistance among the races. Black children and adults show higher skin resistance than do White children and adults (Batson, Young, Shepard, 1962; Johnson, Corah, 1963). Neonates of all races show a high skin resistance, and they also sweat very little. This gives support to the concept that sweat chlorides play a role in galvanic skin response.

The apocrine sweat glands, like the sebaceous glands, open into hair follicles. These glands are located in the axillary, groin, and pubic regions; around the anus, umbilicus, and areola of the breast; and in the external auditory canal. They become functional after puberty. This explains the lack of underarm sweating and body odor in prepubescent children and the acute discomfort of some teenagers concerning axillary sweating: they have not experienced it before.

The number of apocrine glands varies according to race. There are fewer apocrine glands in Orientals and Native Americans than in Blacks and Whites. Evidence that the secretions differ by race may be extrapolated from data on the differences in the composition of dry and wet earwax (cerumen) (Matsunaga, 1962). The alleles for the apocrine glands are involved in earwax type.

Earwax Earwax is produced by the glands in the external ear canal. Unlike the usual apocrine glands, the ceruminous glands start functioning at birth. The peoples of the world can be divided into those with wet earwax and those with dry earwax. Orientals and American Indians have a frequency of dry earwax around 84%; the frequency for this allele in Blacks and Whites is low. The actual frequency of the phenotype for wet earwax is around 97% for Whites and 99% for Blacks. In this two-allele system, wet earwax is dominant over dry wax (Matsunaga, 1962).

There is speculation about the reasons for earwax polymorphism, but no hard data are available. There may be a disease susceptibility difference or an advantage of earwax type by climate. A cline has been found that relates to humidity and temperature: wet wax is common in hot, moist environments (McCullough, Giles, 1970). And a difference has been found in breast cancer susceptibility that correlates with earwax types (Petrakis, 1971). Women with dry earwax have a lower incidence of breast cancer than women with wet earwax. It is postulated that the association of earwax and breast cancer may be caused by the action of a pleiotropic allele that affects the apocrine system. An association between amount of breast secretion and earwax type has also been reported (Petrakis et al, 1975). Women with dry earwax have a lower level of breast secretion than women with wet earwax. Another interesting observation is that White women have larger breasts than Oriental women and that Japanese women who had breast cancer had larger breasts than Japanese noncancer patients (Petrakis, 1971).

Earwax and sweat correlations Earwax type and axillary odor are correlated. Most Orientals and Native Americans have a mild to absent body odor; Blacks and Whites tend to have a strong body odor. Studies on Japanese, Malays, Chinese, and Asian Indians show a strong correlation of body odor with wet earwax (Matsunaga, 1962; Norakmal, Tan, 1979). Over 90% of the individuals with dry earwax do not have body odor, while around 60% of those with wet earwax do have a noticeable body odor. Because this correlation is not perfect, probably another allele is involved or a modifying gene is present at another locus. (I have met White individuals without body odor and have assumed that they had dry earwax. Since I rarely carry an otoscope, I have been unable to verify this. It would be a fascinating study to compare wet earwax individuals with and without body odor to elucidate the genetics of earwax. Unfortunately, the widespread use of deodorants would make this a difficult study.)

The composition of wet and dry wax differs. Wet wax is approximately 50% lipid and 20% protein; the values for dry wax are almost reversed: 18% lipid and 43% protein (Matsunaga, 1962). It may be that axillary odor is caused by the decomposition of the lipids—Blacks and Whites have more lipids, so their odor is stronger. The implications of body odor for judgments concerning personal hygiene are obvious.

Surface Anatomy

A few surface features vary by race and are clinically relevant: lip pits, cleft uvulas, earlobe creases, and fingerprints.

Lip pits Lip pits are located in the commissure of the lips—the crease between the upper and lower lip. They present as little holes. One study measured their depth, which ranged from a slight dimpling to 4 mm. They cause no problem but might be noticed by health workers when inspecting a patient's mouth. Lip pits seem to occur in association with preauricular pits, which almost always go unnoticed. (Preauricular pits are found in front of the ear where the upper auricle originates.) Both lip pits and preauricular pits are normal, not pathological, variations. The frequency of lip pits varies by group. They are found in 12% of Whites, 20% of Blacks, and 7% of Orientals (Schaumann, Peagler, Gorlin, 1970).

Cleft uvulas Cleft uvulas are a startling finding in some individuals; the uvula may be separated at the tip, giving it a fishtail appearance. This is the most common variety, but the separation can range from a split at the tip to complete separation into two uvulas (Meskin, Gorlin, Isaacson, 1964). The condition is rare in Blacks and Whites but reaches a frequency of 10% in Orientals. In some American Indian groups, it can be as high as 18% (Schaumann, Peagler, Gorlin, 1970). There is strong support from family studies that the cleft uvula is a subclinical manifestation of cleft lip and cleft palate. Clefting of the lip and palate is most common in Orientals and American Indians and least common in Blacks, with Whites being intermediate. These figures parallel the frequencies for cleft uvula (Emanuel et al, 1972).

Earlobe creases A deep diagonal crease in the earlobe has come into prominence in the past few years as a sign associated with coronary heart disease in some populations (Frank, 1973). The incidence of the earlobe sign increases with age, beginning around 30, and also seems to be associated with obesity (Kaukola et al, 1979). The sign has caused some controversy because certain populations, specifically the Japanese and Chinese, do not show an association with coronary heart disease (Petrakis, Koo, 1980). A study designed to indicate possible racial variation in the earlobe crease showed that Whites, Blacks, and Latin Americans have an age-related increase in the frequency of the crease, but that Chinese and Japanese do not (Petrakis, Koo, 1980).

A recent study of Eskimos, Navajos, and Whites also showed an age-related increase in earlobe creases (Overfield, Call, 1983). This study found no difference between the groups in the frequency of creasing, but it did find a difference related to age at onset of earlobe creasing. Whites had earlobe creases a decade earlier than Navajos. Earlobe creasing was also found to be related to earlobe type. There are three earlobe types: free, attached, and soldered (Brues, 1977). See Figure 2.1. Free lobes occur most frequently in Blacks and Whites. Orientals have more attached and soldered lobes (Brues,

1977). None of the individuals with soldered lobes had creases; 5–11% of those with attached lobes had creases; and 35% of those with free lobes had creases (Overfield, Call, 1983). The authors concluded that "the ear-lobe crease should be considered more a curiosity than a warning sign of anything but the passage of time" (p 480).

Fingerprints Fingerprints (dermatoglyphics) are largely a hereditary surface variation and are assumed to be a polygenic trait. Fingerprint patterns occur in three basic figures: loops, whorls, and arches. Loops are the most common and arches the least common in all races, but races vary in their proportion of each pattern. Whites and Blacks have approximately 60–65% loops, Orientals and Native Americans about 50%. Whorls range from about 30% for Blacks to 45% for Orientals and Native Americans, with Whites intermediate at 35%. Arches average around 5% for each race (Plato, Cereghino, Steinberg, 1975).

These patterns are formed early in fetal development and are complete by about the fourth month of gestation. During the first trimester, the fetus is extremely susceptible to damage from viral diseases and other environmental influences, such as drugs. This explains some of the variation seen in fingerprint patterns, but it is generally accepted that the patterns are polygenically inherited. An interesting question is whether the differences in patterns seen in chromosomal disease conditions are caused by damage during development or are due to the genetic control of dermatoglyphic patterns as a part of the total disease syndrome.

FIGURE 2.1 Ear lobe types

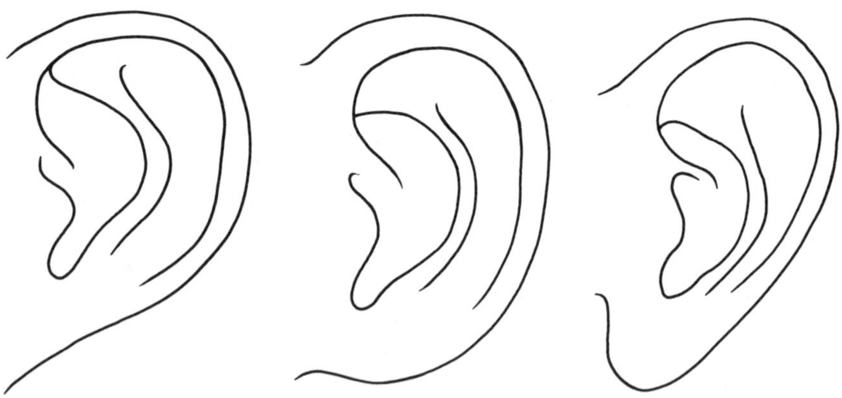

Source: Overfield, Call, 1983, p 479.

A most exciting and pertinent difference in fingerprint patterns is noted in persons with myocardial infarction, compared with normal controls. It has been shown that persons with a history of myocardial infarction have a significantly higher frequency of whorls and correspondingly lower frequency of ulnar loops than the control population (Rashad, Mi, 1975). This study, done on 800 Japanese males between the ages of 55 and 75, also looked at individuals with angina pectoris and hypertension and found that they did not differ from healthy controls. It would be interesting to see a comparable study done on other races, for Orientals have 10–15% fewer whorls than Blacks and Whites.

ANATOMIC VARIATION

Contrasts in body proportion and anatomic differences in bones, teeth, and soft tissues among the races should be known by those providing health care. The following sections present variations that directly affect health care or that health workers can expect to see when giving care to people of other races.

Body Proportions

Height and weight have been extensively studied in Blacks and Whites in the United States (Abraham, Clifford, Najjar, 1976). The mean height of American men 18 to 74 years of age is 175.26 cm (5 ft 9 in); for American women, it is 161.54 cm (5 ft 3.6 in). The figures indicate that White men average 1.27 cm (0.5 in) taller than Black men, but Black and White women are the same height. Black and White men do not differ in weight, but the women do. Black women are consistently heavier at every age than White women; from 35 to 64 years of age, they average 9.1 kg (20 lb) heavier.

Although Black and White men are of similar height and weight, their body proportions differ. Sitting/standing height ratios indicate that Blacks of both sexes have longer legs and shorter trunks than Whites. Because proportionally most of the weight is in the trunk, White men are fatter than Black men. This characteristic is reversed in women: Black women are fatter than White women despite their longer legs.

Bone length, as revealed by stature, shows definite racial differences (Brues, 1977). Blacks tend to have longer legs and arms than Whites. Orientals and American Indians on the average have longer trunks and shorter limbs than Whites. There is a good deal of overlap, of course. Blacks also tend to be wide-shouldered and narrow-hipped, Orientals wide-hipped and narrow-shouldered. Shoulder width is produced, in good part, by the length of the clavicle. Because the clavicle is a long bone, this explains why linearly built individuals have wide shoulders.

It is well known that the children of migrants to this country are taller than their former compatriots. It is also common knowledge that the Japanese in Hawaii are taller than those in Japan. This holds for other groups as well. For instance, Italians in California are 3.8 cm (1.5 in) taller than their former compatriots (Hulse, 1968); however, the sitting/standing height ratio does not appear to change. This indicates that the increase in height occurs in both the long bones and the spine (Brues, 1977).

The height increase in migrants is thought to result from better nutrition and decreased interference with growth from infectious diseases during the formative years. These same conditions probably account for the secular trend in height that is generally taking place in this country. The average height of American men has increased 1.8 cm (0.7 in) in 10 years, while that of women has increased 1.3 cm (0.5 in) in the same period (Abraham, Clifford, Najjar, 1976).

One area where people differ markedly in body size and shape is in the amount of body fat. By far the best predictor of obesity is socioeconomic status (Goldblatt, Moore, Stunkard, 1965). The lower class is fatter than the middle class, which in turn is fatter than the upper class. As individuals rise in socioeconomic circumstances, they tend toward normal-to-thin weight.

There are also racial differences in the amount of body fat, practically all of which are related to economic status. In addition, national differences, again mostly because of socioeconomic class, affect the characteristics of body fat. Some evidence indicates that different races put on fat differently. Blacks tend to have smaller skinfold thicknesses in their arms than Whites; the distribution of fat on the trunk is similar (Robson, Bazen, Soderstrom, 1971). These differences are statistically significant but not clinically relevant because, on the average, only 1 mm of fat separates the two groups.

A confounding factor in all studies of racial differences in body fat is climate. Peoples of all races tend to have more body fat in cold climates and less in warm climates (Newman, 1956). Thus Black-White differences are probably at least partially related to climate.

Another common observation is that individuals within all races put on weight in different parts of the body: big tummies, saddle-bag thighs, large derrieres, beer bellies, and pear shapes. These spot-fat deposits are all harbingers of a more general deposition of fat, if they are ignored. Unfortunately for people with spot fat, so-called spot-reducing exercises appear to be ineffective (Gwinup, Chelvam, Steinberg, 1971). General dieting and exercise are needed to decrease fat deposits.

Socioeconomic status is an important variable in obesity for two major reasons: malnutrition is associated with socioeconomic status and social pressure against obesity differs by socioeconomic status. Several studies done in the United States on large numbers of subjects have shown that an individual's social status has a major effect on obesity. A study of 1,000 women indicates that 30% of lower-class women, 16% of middle-class women, and 5% of upper-class women are obese (Goldblatt, Moore, Stunkard, 1965). Another study of 1,500 girls presents similar results (Stunkard et al, 1972). Many other studies have confirmed these findings.

Social mobility also contributes to weight gains and losses. Among people in a higher social class than their parents, only 12% are obese, while among those who have moved down in social class 22% are obese (Goldblatt, Moore, Stunkard, 1965). Closely related to social class is the number of generations a family has been in this country; 24% of first-generation women are overweight, compared with only 5% of fourth-generation women. Obesity in men follows the same trends but is less marked.

The implications of the influence of social class on obesity are important. Obesity cannot be viewed as an abnormal characteristic of individuals in the lower class. To the layperson *abnormal* means something out of the ordinary. If 1 out of every 3 or 4 women in the neighborhood is fat, then obesity is not abnormal. But if only 1 out of 20 is fat (as in the upper class) or only 3 out of 20 are fat (as in the middle class), that becomes abnormal.

Two recent studies give some insight into the reason obesity is prevalent in the lower classes. One study, of 300,000 men examined before induction into the Dutch Army (Ravelli, Stein, Susser, 1976), is extremely exciting in its implication for prevention. These young men were either in utero or infants during World War II. In the Netherlands, a 7-month famine was caused when the Germans placed an embargo on food as a reprisal for Dutch resistance. The data derived from this famine were used to examine the effects of malnutrition on all phases of gestation and early development. The study indicates that inadequate nutrition during the last trimester of pregnancy and the first month of life produces significantly lower rates of obesity in the offspring than were found in the controls. Nutritional deprivation affects a critical period in the development of adipose tissue. But what is so interesting about this study is its indication that malnutrition occurring in the first half of pregnancy results in offspring with significantly higher rates of obesity compared with controls. The hypothalamic center in the brain, which regulates appetite and hence food intake, develops during the first half of pregnancy. Malnutrition at this critical time appears to affect the hypothalamus, causing a lifelong overeating problem.

A similar study of Danish inductees corroborates the findings of the Dutch study (Sonne-Holm, Sorensen, 1977).

Bones

The long bones of Blacks are significantly longer than those of Whites and also significantly narrower (Farrally, Moore, 1975). A long, narrow bone would have less tensile strength than a shorter, wider bone unless its substance was denser, as is the case in Blacks (Trotter, Peterson, 1966). A study of 200 skeletons from 100 Blacks and 100 Whites revealed that Blacks have denser bones than Whites; men have denser bones than women; and the density of bone decreases with age in both sexes and races at a uniform and parallel rate (Trotter, Peterson, 1966).

This bone density difference by race and sex explains why women have a higher incidence of osteoporosis and bone fractures in later years than men do. It also explains why osteoporosis is a relatively rare disease in Blacks. On a scale of bone density, Black men have the densest bones, followed by Black women and White men, with similar densities; White women have the least dense bones. Data indicate that bone density in the Chinese, Japanese, and Eskimos is below that of White Americans (Garn, 1964; Mazess, Mather, 1975).

Curvature of the long bones is another characteristic that differs by race (Gilbert, 1976). American Indians have strongly anteriorly convex femurs; American Blacks have markedly straight femurs; American Whites are intermediate. This characteristic is thought to be genetically determined to some extent, but curvature also seems to be influenced by body weight. Thin Blacks and Whites have less curvature than the means for their respective groups; obese Blacks and Whites exhibit more curvature than the mean. It is possible that the heavier density of the bones of Blacks helps to protect them from curvature due to increased weight bearing. Several theories regarding femoral curvature have been disproved: it is not caused by being an equestrian nomad; it is not a pathologic condition; and it is not due to postural habits, such as squatting (Gilbert, 1976).

The degree of torsion, or rotation, of the proximal end of the humerus in response to muscle pull differs by race. Whites show more humeral torsion than Blacks, and it is more marked on the right side than the left (Krahl, 1976). When Blacks do have humeral torsion, it is bilaterally symmetrical. This is probably another response of the bones to differences in density—the more dense bones being less susceptible to torsion from muscle activity.

There are race and sex differences in skull thickness that do not seem to be related to the rest of the skeleton (Adeloye, Kattan, Silverman, 1975). X-ray films of the skulls of 500 Blacks and Whites indicate that the frontal bones of White men are thicker than those of Black men; the opposite is true of the parietal and occipital bones. It is commonly believed that men have thicker skulls than women; however, at some ages and in some areas of the skull, women have significantly thicker skulls than men. For example, after age 60, women have thicker frontal bones than men. Age changes are similar for both races and sexes. The skull rapidly increases in thickness in the first two decades and then increases more slowly from the third to the sixth decade.

An oddity of bone length differences is the variable nature of the length of the second toe. It is an inherited, most probably polygenic, trait. In most people the big toe is the longest one. Populations differ in the frequency of longer second toes. There are some racial trends, but there is an even greater spread among different populations. In White groups, those with longer second toes range from 8% to 34%. In Blacks, they range from 8% to 12%. Vietnamese have a 31% frequency, and Melanesians, a 21–57% frequency (Papadopoulos, Damon, 1973). Little work has been done on this variation, despite the problems that longer second toes apparently give joggers.

Another bony variation concerns the radius and ulna: they can be the same length at the wrist (neutral arrangement), or either bone can protrude farther than the other (Chan, Huang, 1971). This possibly insignificant finding differs by race in the two populations studied, the Swedes and the Chinese. The bones were of equal length at the wrist in 61% of Swedes and 42% of Chinese. The ulna was longer than the radius in 16% of Swedes and 48% of Chinese. A longer radius was present in 23% of the Swedes and 10% of the Chinese.

A more interesting variation in bone is differences in the number of vertebrae. The usual number, 24 vertebrae, is present in 85–93% of individuals (Kaufman, 1974); 3–11% have 23 vertebrae, and 3–8% have 25 vertebrae.

The usual pattern for 24 vertebrae is 7 cervical (C7), 12 thoracic (T12), and 5 lumbar (L5), but variations do occur in the proportions of thoracic and lumbar vertebrae. Two common variations are T11, L6, and T13, L4. For 23 vertebrae, the common patterns are T11, L5, and T12, L4. For 25 vertebrae, the common patterns are T12, L6, and T13, L5.

There are differences by race and sex in the number of vertebrae: women have 23 vertebrae more frequently than men. The differences by race tend to blur because of sex differences, but 11% of American Black females have 23 vertebrae—the most frequent occurrence of this variation. The 25-vertebrae variant occurs most frequently among Eskimo and Indian males, 12% of whom are affected. Other percentages by race are displayed in Table 2.1.

There may be a correlation between the number of vertebrae and lower back pain. My guess is that individuals with fewer than the average number of lumbar vertebrae have less lordosis and suffer less back trouble, while those with an increased number of lumbar vertebrae have a higher incidence of lordosis and back trouble. Because individuals with 23 vertebrae

TABLE 2.1 Percentage distribution of number of vertebrae by race and sex

Racial group	Usual number 24		Variant numbers 23		Variant numbers 25	
	M	F	M	F	M	F
American Blacks	91	87	5	11	4	2
American Whites	90	90	3	6	7	4
Eskimos and Indians	85	93	3	3	12	4
Japanese	91	90	4	5	5	5

Source: Adapted from Kaufman, 1974.

would have 5 or fewer lumbar vertebrae, such individuals should be underrepresented in the population experiencing lower back pain.

Bony protuberances in the mouth, called tori, are noticeable features in some groups. Mandibular tori occur on the lingual surface of the mandible, usually near the canine and premolar teeth. Palatine tori occur along the suture line of the hard palate. Evidence suggests that tori are inherited as an autosomal dominant trait. Except for the fact that they cause fitting problems for denture wearers, tori are harmless.

Bilateral or unilateral mandibular tori are present in approximately 8% of Blacks and Whites (Axelson, Hedegard, 1981). They are more common in the Oriental and Native American races; the prevalence rate varies by specific subgroup, exceeding 50% in some Eskimo groups (Jarvis, Gorlin, 1972).

Palatine tori occur in around 24% of Whites, 20% of Blacks, and a variable but generally higher percentage of Native Americans and Orientals. Up to 50% of them have palatine tori (Schaumann, Peagler, Gorlin, 1970; Jarvis, Gorlin, 1972).

Pelvic Measurements and Newborn Size

Pelvic dimensions, the size of the newborn, and the length of gestation differ according to race. The interaction among these factors allows women with different body builds to give birth safely. Adjustments among the factors evolved over time—where there was a disparity, a woman died during childbirth, reducing the instances of passing on her genes to the next generation.

There is a well-documented difference in birth weight between Blacks and Whites. On the average, Black infants weight 240 g (8.5 oz) less than

White infants (National Center for Health Statistics, 1980). Prematurity estimates based on weight now take this difference into consideration (Mellits, Hardy, 1977). The revised lower estimate of prematurity weight is based on birth weight and assessment of maturity. Size for size, Black infants are generally more mature than White infants (Mellits, Hardy, 1977).

The weight differences are caused partially by a gestation period 9 days shorter for Blacks and partially by the slowing of intrauterine growth of Black infants after 35 weeks of gestation. Before 35 weeks, Black infants are usually larger than White infants (Pratt, Janus, Saval, 1977). Why intrauterine growth slows and why gestation is terminated earlier are not known.

Along with the smaller size of Black infants goes a smaller head circumference. Black infants have on the average a 0.7 cm (0.25 in) smaller head than White infants. The smaller head circumference, birth weight, and gestational length in Black infants, compared to Whites, could all be explained as adaptations to the smaller pelvises of Black women. Unfortunately, data to substantiate this are scanty and indirect at best. Black women do have a 1.6 cm (0.6 in) smaller hip width than white women (Meredith, 1971). There is also a difference in the percentages of pelvic types. Both races have approximately 42% gynecoid pelvis types, but White women have more android pelvic types, while Black women have more anthropoid pelvic types (Williams, Warwick, 1980). Because hip width is a general measure of size, perhaps the type of pelvis is less important than its general size. A small gynecoid pelvis might be the cause of more dystocia than a larger android or anthropoid type. I found no exact pelvic measurements to refute or substantiate this notion, but such a study would be rather simple to do.

Chest Size and Pulmonary Function

Pulmonary function, as determined by vital capacity and forced expiratory volume, differs significantly by race (Oscherwitz et al, 1972; Lapp et al, 1974). We may reasonably assume that such differences are related to the size of the chest cavity. Whites have the largest chest volumes, followed by Blacks, then Orientals and Native Americans. Although Blacks and Whites are similar in height, chest volume is smaller in Blacks. When the shorter height of Orientals is taken into consideration, their chest volume is still significantly lower than in either Blacks or Whites (Oscherwitz et al, 1972).

These lung volume differences are great enough to be taken into consideration when using the following pulmonary function tests: forced expiratory volume (FEV), forced vital capacity (FVC), and the derived function FEV/FVC%. Table 2.2 shows the differences among the three races in these measurements, which were taken on healthy men between 38 and 70 years of age (Oscherwitz et al, 1972).

TABLE 2.2 Pulmonary function results by race

Race	Number	FEV (liters)	FVC (liters)	FEV/FVC%
Whites	465	3.22	4.30	74.4
Blacks	98	2.85	3.70	76.7
Orientals and Native Americans	59	2.53	3.27	77.0

Source: Adapted from Oscherwitz et al, 1972, p 323.

Teeth

Teeth are obvious structures that display much variation according to race. Races differ in tooth size, shape, and number, as well as in the amount of tooth decay and periodontal disease. The variations in tooth disease susceptibility are discussed in chapter 4. The rate at which teeth erupt in growing children also differs by race; this is covered in chapter 3.

Because teeth are often used as indicators of developmental, hygienic, and nutritional adequacy, health care providers should note the differences that occur by race. Size differences are probably the most apparent, especially when individuals of more than one race are seen over a short span of time. The teeth of Whites are the smallest of any race, followed by Blacks, then Orientals and Native Americans. Australian Aborigines and some Eskimos have the largest teeth. In general, Native Americans have somewhat larger teeth than Orientals (Dahlberg, 1963).

Tooth-size differences among races help explain the differential occurrence of prognathism (protruding jaws). Tooth size and jaw size are positively correlated (Anderson, Thompson, Popovich, 1977). Hence prognathism is seen more frequently in Blacks and Orientals.

Teeth vary in shape as well as size. The incisors of Orientals and American Indians are most often shovel-shaped, the lingual side of the teeth having a scooped-out configuration. Blacks and Whites show this feature only occasionally (Carbonell, 1963). The variation is polygenetically inherited. The reasons for its occurrence remain obscure, but teeth with this shape are heavier and can probably withstand more wear.

A conspicuous difference in tooth shape is a peg tooth—a small vestigial tooth. The most frequently pegged teeth are lateral incisors. These occur in 5–8% of Orientals and 1–4% of Blacks and Whites (Escobar, Conneally, Lopez, 1977). Pegged teeth often occur in the same individuals who lack certain other teeth, especially the third molar (LeBot, Salmon, 1977).

A trend toward fewer teeth is occurring in the human race in general. The average adult number of teeth is 32, but a reduction to 28 is common. The most commonly missing teeth are the third molars (wisdom teeth). The next most common are the lateral incisors (Brothwell, Carbonell, Goose, 1963). A pegged lateral incisor is considered a generation or two away from complete absence. Thus a pegged tooth in one generation may be a predictor of agenesis in the following generation. Contrary to the trend, Australian Aborigines and Melanesians may have up to four extra molars, resulting in a very large mouthful of teeth (Brues, 1977).

Agenesis itself varies by race. Orientals have the most third molar agenesis, occurring in 18–35% of the population. Whites show the trait less often, at 9–25%, and Blacks the least, at 1–11% (Brothwell, Carbonell, Goose, 1963).

Soft Tissues

Soft tissue differences occur in blood vessels and muscles. There are many different blood vessel patterns that vary by race, but I will describe only one: the pattern of superficial veins on the anterior chest. There are also several muscle variations of which I will describe two: the peroneus tertius and palmaris longus muscles.

Mammary venous plexus

There are two patterns of venous drainage in the mammary region, called the *transverse* and the *longitudinal* patterns. They occur in both sexes and appear to be controlled by a single dominant gene (Spuhler, 1950). The only known alterations of either vein pattern are produced by breast tumors. The superficial veins of the chest form a network over the entire chest, of which the portion over the mammary glands is called the *mammary venous plexus*. In the transverse pattern, the veins in this plexus radiate laterally toward the axillae. In the longitudinal pattern the veins radiate downward and laterally like a fan. Figure 2.2 shows the two patterns schematically. To classify pattern, a health professional looks at the veins over the mammary glands, not the upper chest, because the patterns are sometimes different there.

The veins are more or less noticeable in both men and women. During pregnancy they are more distinct; the pattern reaches maximum clarity when lactation begins.

These two patterns occur with different frequencies in the two populations that have been studied. White women have the recessive longitudinal pattern 6–10% of the time; in Navajos, the longitudinal pattern occurs 30% of the time. The gene frequencies for the two populations are 0.24–0.32 for the recessive gene in Whites, and 0.55 in Navajos.

Examination of the mammary venous plexus in pregnant women or, under appropriate lighting, in all women could become part of the general physical exam if more data were available on their distribution. It is surprising that these vein patterns have not been studied in other populations, because the only known alteration of either pattern is produced by breast tumor (Spuhler, 1950).

Muscles The peroneus tertius and palmaris longus muscles are absent in some individuals. The peroneus tertius muscle flexes the foot upward (dorsiflexion). Since the tibialis anterior also does this, lack of the muscle causes no problems. The peroneus tertius inserts at the base of the fifth metatarsal and has its origin on the front of the lower third of the fibula. When a person stands with toes in sharp dorsiflexion, the tendon of the peroneus tertius can be felt over the cuboid bone in the ankle (Spuhler, 1950). In Blacks, this muscle is absent most often, 10–15% of the time, and in Orientals, Native Americans, and Whites least often, 3–10% of the time. The inheritance of this condition is not well understood. Individuals may even have the muscle on one side but not the other. Females lack the muscle more often than males, and there are great population differences. For instance, 24% of Berbers, from the Sahara desert, lack the muscle.

The palmaris longus muscle is involved in wrist flexion; since three other muscles are also used for flexion, absence of the palmaris longus does not cause problems. This muscle inserts on the transverse carpal (wrist ligament) and originates on the medial epicondyle of the humerus. Some 12–20% of Whites lack it, as do approximately 5% of Blacks, 3% of Orientals, and 2–19% of American Indians (Reinmann et al, 1944).

FIGURE 2.2 Venous drainage patterns in the mammary region

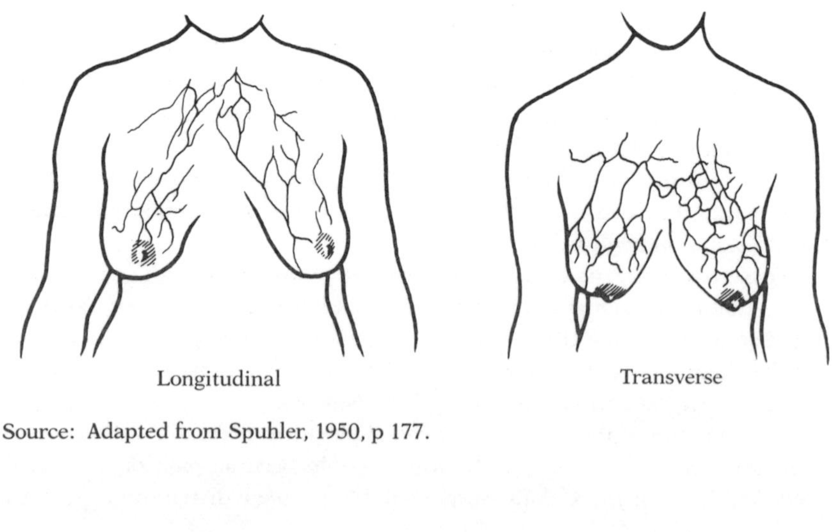

Source: Adapted from Spuhler, 1950, p 177.

Summary

This chapter on surface and anatomic variations among the races is an introduction to the great amount of variation that occurs internally. In this and the following chapters, some of the variations have specific health implications, while others are singled out to help distinguish normal variation from pathology or peculiarity.

Surface and anatomic variations with health care implications include retinal pigmentation, earlobe creases, bone density, vertebrae number, pulmonary function, and venous drainage patterns in the mammary plexus. In addition, health care providers must be able to distinguish mongolian spots from bruises and should know that there is no correlation between breast pigmentation and breast-feeding difficulties.

Other racial variations are presented simply to inform health workers about the variety that they can expect to see when giving care to people of other races. These differences include the influence of the environment on the number of eccrine sweat glands, sweat chloride levels, and galvanic skin resistance, earwax and body odor correlations, incidence of lip pits and cleft uvulas, pelvic measurement and newborn size differences by race, and variability in the number and shape of the teeth.

REFERENCES

Abraham SJ, Clifford L, Najjar MF: Height and weight of adults 18–74 years of age in the United States. *Advancedata* (November 19) 1976; (3):1–8.

Adeloye A, Kattan KR, Silverman F: Thickness of the normal skull in the American Blacks and Whites. *Am J Phys Anthropol* 1975; 43(1):23–30.

Anderson DL, Thompson GW, Popovich F: Tooth, chin, bone and body size correlations. *Am J Phys Anthropol* 1977; 46(1):7–11.

Axelson G, Hedegard B: Torus mandibularis among Icelanders. *Am J Phys Anthropol* 1981; 54(3):383–389.

Batson R, Young WC, Shepard FM: Observations on skin resistance to electricity and sweat chloride content. *J Pediatr* 1962; 60(5):716–720.

Bojsen-Moller F: Extensor carpi radialis longus muscle and the evaluation of the first intermetacarpal ligament. *Am J Phys Anthropol* 1978; 48(2):177–184.

Bor S, Feiwel M, Chanarin I: Vitiligo and its aetiological relationship to organ-specific autoimmune disease. *Br J Dermatol* 1969; 81:83–88.

Brothwell DR, Carbonell VM, Goose DH: Congenital absence of teeth in human populations. Pages 179–189 in: *Dental Anthropology*. Brothwell DR (editor). Pergamon Press, 1963.

Brown MS, Hurlock JT: Preparation of the breast for breastfeeding. *Nurs Res* 1975; 24(6):448–451.

Brues AM: *People and Races*. Macmillan, 1977.

Brues AM: Rethinking human pigmentation. *Am J Phys Anthropol* 1975; 43(3):387–391.

Carbonell VM: Variations in the frequency of shovel-shaped incisors in different populations. Pages 211–233 in: *Dental Anthropology*. Brothwell DR (editor). Pergamon Press, 1963.

Chan KP, Huang P: Anatomic variations in radial and ulnar lengths in the wrists of Chinese. *Clin Orthop* 1971; 80:17–21.

Dahlberg AO: Analyses of the American Indian dentition. Pages 149–177 in: *Dental Anthropology*. Brothwell Dr (editor). Pergamon Press, 1963.

Emanuel I et al: The incidence of congenital malformations in a Chinese population: The Taipei Collaborative Study. *Teratology* 1972; 5(2):159–170.

Escobar V, Conneally PM, Lopez C: The dentition of the Queckchi Indians. *Am J Phys Anthropol* 1977; 47(3):443–451.

Farrally MR, Moore WJ: Anatomical differences in the femur and tibia between Negroes and Caucasians and their effects on locomotion. *Am J Phys Anthropol* 1975; 43(1):63–69.

Frank ST: Aural sign of coronary artery disease. *N Engl J Med* 1973; 289(6):327–328.

Gans B: Breast and nipple pain in early stages of lactation. *Br Med J* 1958; 5100:830–832.

Garcia RI et al: Number of epidermal melanocytes, hair follicles and sweat ducts in skin of Solomon Islanders. *Am J Phys Anthropol* 1977; 47(3):427–433.

Garn SM: Compact bone in Chinese and Japanese. *Science* 1964; 143(3613):1439.

Gilbert BM: Anterior femoral curvature: Its probable basis and utility as a criterion of racial assessment. *Am J Phys Anthropol* 1976; 45(3):601–604.

Goldblatt P, Moore M, Stunkard A: Social factors in obesity. *JAMA* 1965; 192:97–102.

Goodman RM: *Genetic Disorders Among the Jewish People*. Johns Hopkins University Press, 1979.

Gwinup G, Chelvam R, Steinberg T: Thickness of subcutaneous fat and activity of underlying muscles. *Ann Intern Med* 1971; 74:408–411.

Hoffman MJ: Retinal pigmentation, visual acuity and brightness levels. *Am J Phys Anthropol* 1975; 43(3):417–424.

Hulse FS: The breakdown of isolates and hybrid vigor among the Italian Swiss. *Proc 12th Intl Cong Genet* 1968; 2:177.

Jacobs AH, Walton RG: Incidence of birthmarks in the neonate. *Pediatrics* 1976; 58:218–222.

Jarvis A, Gorlin RJ: Minor orofacial abnormalities in an Eskimo population. *Oral Surg* 1972; 33(3):417–427.

Johnson LC, Corah NL: Racial differences in skin resistance. *Science* 1963; 139(3556):766–767.

Kaufman PDB: Variation in the number of presacral vertebrae in Bantu-speaking South African Negroes. *Am J Phys Anthropol* 1974; 40(3):369–374.

Kaukola A et al: Ear lobe crease and coronary atherosclerosis. *Lancet* 1979; 8156/7(II):1377.

Krahl VE: The phylogeny and ontogeny of humeral torsion. *Am J Phys Anthropol* 1976; 45(3):595–599.

Kuno Y: *Human Perspiration.* Thomas, 1956.

Lapp NL et al: Lung volumes and flow rates in Black and White subjects. *Thorax* 1974; 29:185–188.

LeBot P, Salmon D: Congenital defects of the upper lateral incisors. *Am J Phys Anthropol* 1977; 46(2):231–243.

Levin S et al: Effect of age, ethnic background and disease on sweat chlorides. *Isr J Med Sci* 1966; 2(3):333–337.

Martin JL, Crump EP: Leukoedema of the buccal mucosa in Negro children and youth. *Oral Surg* 1972; 34(1):49–58.

Matsunaga E: The dimorphism in human normal cerumen. *Ann Hum Genet* 1962; 25:273–286.

Mazess RB, Mather WE: Bone mineral content in Canadian Eskimos. *Hum Biol* 1975; 57(1):45–63.

McCullough JM, Giles E: Human cerumen types in Mexico and New Guinea: A humidity-related polymorphism in Mongoloid peoples. *Nature* 1970; 226(5244):460–462.

Mellits ED, Hardy JB: Information pertaining to birth weight, gestational age, placental weight and mortality. Pages 119–140 in: *The Epidemiology of Prematurity.* Reed DM, Stanley FJ (editors). Urban & Schwarzenberg, 1977.

Meredith HV: Worldwide somatic comparisions among contemporary human groups of adult females. *Am J Phys Anthropol* 1971; 34(1):89–132.

Meskin LH, Gorlin RJ, Isaacson RJ: Abnormal morphology of the soft palate: The prevalence of cleft uvula. *Cleft Palate J* 1964; 1:342–346.

National Center for Health Statistics: Final natality statistics, 1978. *Monthly Vital Stat Rep* 1980; 29(1):1–27.

Newman RW: Skinfold measurements in young American males. *Hum Biol* 1956; 28:154–164.

Nicolaides N, Rothman S: Studies on the chemical composition of human hair fat. *J Invest Dermatol* 1953; 21:9–14.

Norakmal I, Tan SG: Cerumen polymorphism in the three major ethnic groups of Malaysia. *Jpn J Hum Genet* 1979; 24:119–121.

Oscherwitz et al: Differences in pulmonary functions in various racial groups. *Am J Epidemiol* 1972; 96(5):319–327.

Overfield T, Call EB: Earlobe type, race, and age: Effects on earlobe creasing. *J Am Geriatr Soc* 1983; 31(8):479–481.

Papadopoulos CC, Damon A: Some genetic traits in Solomon Island populations: III. Relative toe length. *Am J Phys Anthropol* 1973; 39(2):185–189.

Pawson EG, Petrakis NL: Comparisons of breast pigmentation among women of different racial groups. *Hum Biol* 1975; 47(4):441–450.

Petrakis NL: Cerumen genetics and human breast cancer. *Science* 1971; 173:347–349.

Petrakis NL, Koo L: Earlobe crease. *Lancet* 1980; 8164(I):376.

Petrakis NL et al: Association of race, age, menopausal status and cerumen type with breast fluid secretion in non-lactating women as determined by nipple aspiration. *J Natl Cancer Inst* 1975; 54(4):829–833.

Plato CC, Cereghino JJ, Steinberg FS: The dermatoglyphics of American Caucasians. *Am J Phys Anthropol* 1975; 42(2):195–210.

Pratt MW, Janus ZL, Saval NC: National variations in prematurity (1973 and 1974). Pages 53–79 in: *The Epidemiology of Prematurity*. Reed DM, Stanley FJ (editors). Urban & Schwarzenberg, 1977.

Rashad MN, Mi MP: Dermatoglyphic traits in patients with cardiovascular disorders. *Am J Phys Anthropol* 1975; 42(2):281–283.

Ravelli G, Stein Z, Susser M: Obesity in young men after famine exposure *in utero* and early infancy. *N Engl J Med* 1976; 295(7): 349–353.

Reinmann AF et al: The palmaris longus muscle and tendon: A study of 1600 extremities. *Anat Rec* 1944; 89:495–505.

Resnik S: Melasma induced by oral contraceptive drugs. *JAMA* 1967; 199(9):601–605.

Robson JRK, Bazen M, Soderstrom R: Ethnic differences in skin fold thickness. *Am J Clin Nutr* 1971; 24:864–868.

Schaefer O et al: Regional sweating in Eskimos compared with Caucasians. *Can J Physiol Pharmacol* 1974; 52(5):960–965.

Schaumann BF, Peagler FD, Gorlin RJ: Minor orofacial anomalies among a Negro population. *Oral Surg* 1970; 29(4):566–575.

Sonne-Holm S, Sorensen TI: Post-war course of the prevalence of extreme overweight among Danish young men. *J Chronic Dis* 1977; 30:351–358.

Spuhler JN: Genetics of three normal morphological variations: Patterns of superficial veins of the anterior thorax, peroneus tertius muscle and number of vallate papillae. *Cold Spring Harbor Symp Quant Biol* 1950; 15:175–189.

Stunkard A et al: Influence of social class on obesity and thinness in children. *JAMA* 1972; 221(6):579–584.

Szabo G: The human skin as an adaptive organ. Pages 39–58 in: *Physiological Anthropology.* Damon A (editor). Oxford University Press, 1975.

Szabo G et al: Racial differences in the fate of melanosomes in human epidermis. *Nature* 1969; 222:1081–1082.

Trotter M, Peterson RR: Some variable factors in the adult skeleton. *Ann NY Acad Sci* 1966; 134(2):841–845.

Wasserman HP: *Ethnic Pigmentation: Historical, Physiological and Chemical Aspects.* American Elsevier, 1974.

Williams PL, Warwick R (editors): *Gray's Anatomy.* 36th ed. Pages 386–389. Saunders, 1980.

Woolf CM, Dukepoo FC: Hopi Indians, inbreeding and albinism. *Science* 1969; 164:30–37.

chapter 3

Developmental variation: gestation to geriatrics

Races differ in their rates of growth and development. They also differ in indicators of aging and in age at death. Many of these differences are well documented, but the reasons for them are not completely understood. Both genetic and environmental factors are implicated, the latter including climate, altitude, diet, disease, and socioeconomic conditions.

This chapter examines racial differences in development and their probable origins. Readers will have to sort out the likely causes of these differences in the particular populations that interest them.

Differences in growth and development have been more thoroughly studied than aging differences and are probably more important to the clinician. The factors involved in early growth and development can be modified more easily than the habits or habitats of older persons. Growth and developmental variations before adulthood are presented first. Aging differences follow.

GROWTH AND DEVELOPMENT

Racial differences in growth and development encompass body proportions, developmental maturity, surface features, clinical measurements, and disease susceptibility. Most of these differences have a genetic component, although environmental influences (most significantly climate, altitude, diet, and disease) may amplify or decrease the genetic effects.

Body Size and Proportion Differences

Height and growth rate   Numerous studies have shown the heritability of stature. Tall parents tend to produce tall children; parents of disparate heights produce children of intermediate height. Three kinds of factors seem to be involved in stature: some determine the potential height the individual is capable of reaching, others determine the rate at which full size is achieved, and still others affect the length of time available for growth. Each of these factors can be affected by the environment. For example, poor nutrition may stunt growth by slowing its rate, but it can also delay puberty, thus increasing the time available for growth. Serious illness during a growth period also has a growth-retarding effect.

Growth retardation can be determined by radiography of the long bones. When growth is impeded, dense deposits of calcium are laid down in the growth zone. These lines are called *Harris lines*, after their discoverer. When growth restarts, a line of dense bone is visible on x-ray film. Harris lines correlate well with disease experience, as well as with other insults to growth: malnutrition or starvation, immunizations, and minor surgery (Blanco et al, 1974).

A study of Guatemalan children has shown that reduced stature correlates with the presence of Harris lines (Blanco et al, 1974). Guatemalan children 1 to 7 years of age who had Harris lines were shorter than those without such lines. In boys, the mean height difference was 2.2 cm (0.88 in). Under the same conditions of deprivation, girls show fewer Harris lines than do boys, and their height is less impaired than that of boys.

Compensatory growth is possible after a period of stunting caused by nutritional deprivation or disease, provided the conditions that caused the growth retardation are not prolonged (Ashworth, 1969; Barr, Allen, Shinefield, 1972). Children relieved of the condition that slowed their growth increase their rate of growth and frequently equal their peers in height. However, in the case of chronic malnutrition or disease, accelerated growth may never occur, and the children may never reach their potential in stature (Martorell et al, 1979).

Although it is difficult to separate nongenetic from genetic influences, some populations are shorter or taller than others. In general, the races vary in height from tall to short with Black and White children leading; American Indians are next, being similar to Blacks and Whites or somewhat shorter; then come Mexican-American and Asian children, who are the shortest (Heller, Scott, Hammes, 1967; Barr, Allen, Shinefield, 1972; Yanocheck-Owen, White, 1977; Foster et al, 1977; Johnston et al, 1978; Duncan, Smith, Briese, 1979). In all groups children of higher socioeconomic status are taller (Johnston et al, 1978; Schutte, 1980).

It is postulated that these differences in height are due to a selective advantage developed over many generations. One illuminating Peruvian study on body size showed that, under poor socioeconomic conditions, parents of small body size had a significantly better rate of offspring survival than larger parents (Frisancho et al, 1973).

Obese children are taller than lean children from the same population. A study of low-income White children demonstrates this fact (Garn, Clark, Guire, 1974). Approximately 750 obese children (in the 85th percentile and above for skinfold measurements) and 750 lean children (in the 15th percentile and below) were compared for height differential. The obese children were as much as 6 cm (2.38 in) taller than the lean children. The height differences increased with age, until at 11 to 12 years the obese children were approximately one standard deviation taller than the lean ones. While getting too fat, they were getting enough nutrients to achieve optimum growth—but at a price.

Data on Black and White children between 6 and 11 years show that at age 6 the Black children are taller than the White children. Around age 9 or 10 the White boys begin to catch up in height. White girls do not catch up with Black girls until around 14 or 15 years (Malina, Hamill, Lemeshow, 1974; Hamill, Johnston, Lemeshow, 1973). Around puberty Black children begin to slow down in growth, and White children catch up to them, so that the two races achieve similar heights. However, their proportions (mentioned in chapter 2) and maturation rates still differ. Irrespective of socioeconomic status, Blacks are skeletally more mature than Whites, and girls more mature than boys, at a given age. Chinese and Japanese children raised in this country are delayed in skeletal maturation rate compared to White children, while Oriental children raised in China and Japan are delayed compared to their cohorts in this country (Roche, Roberts, Hamill, 1978). These racial differences are influenced by both genetics and environment.

Birth weight In the United States, Blacks and Whites differ in mean birth weight, with Blacks approximately 240 g (8.5 oz) lighter (National Center for Health Statistics, 1981). This helps to explain why prematurity, defined as birth weight under 2,500 g (5 lb 8 oz), is twice as common in Blacks as it is in Whites. When comparisons are made that control for income, maternal age, parity, and smoking, Blacks still weigh less than Whites at birth. They are also around 2 cm (0.78 in) shorter (Pratt, Janus, Sayal, 1977).

Other races also have lower mean birth weights than Whites: Puerto Ricans, Japanese, Chinese, and Filipinos (Morton, 1977). It has been suggested that the definition of prematurity should be lower than 2,500 g for these groups: Black and Filipino prematurity redefined to 2,200 g, Chinese and Japanese to 2,300, and Puerto Rican to 2,400 (Morton, 1977).

Intertribal variation in birth weight among Native Americans is as great as that between any two major races (Adams, Niswander, 1973). The mean birth weight among Hopi females is 3,097 g (6 lb 13 oz). Among Cheyenne females it is 3,459 g (7 lb 10 oz)—a difference of 362 g (13 oz).

A study done on Blacks in Philadelphia shows something that has been known intuitively for years. The mean birth weight is close to the optimum birth weight for survival (Ellis, 1972). Optimum and mean birth weights for the Philadelphia Black infants were 3.5 kg and 3.0 kg, respectively (7.7 and 6.6 lb). Mortality was lowest in this range, as the curves in Figure 3.1 show. There is higher mortality at the lower birth weights than at the higher weights, but both extremes are risky.

Note also that the mean and optimum birth weights differ in Figure 3.1. This difference is not unique to Philadelphia Black infants; several other studies demonstrate the same discrepancy (Ellis, 1972). The discrepancy is due to excess of less (or greater) than optimal birth weights. Movement of the mean toward the optimal birth weight will be slow even though optimal weight survival is close to 99%. This is because general survival at all other birth weights combined is a little over 95%. A 4% difference is not enough to produce a rapid increase in average birth weight.

Associated with differences in birth weight are differences in body proportions, specifically head and chest circumferences, at birth. The 0.7 cm

FIGURE 3.1 Relationship between birth weight and mortality rate

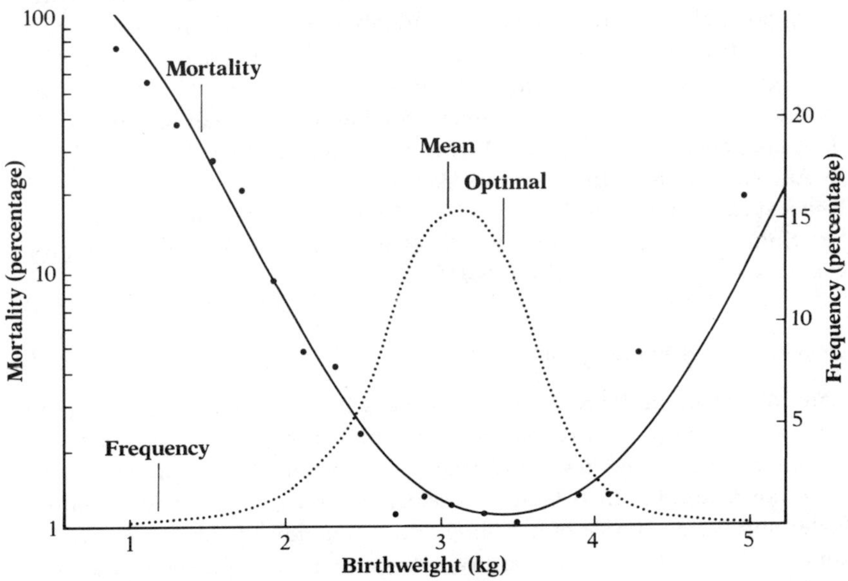

Source: Ellis, 1973, p 147.

(0.25 in) smaller head circumference of Black newborns compared with Whites was mentioned in chapter 2 in relation to maternal pelvic measurements. The average head circumference at birth for populations around the world is 16.4–20.2 cm (12.5–14 in) (Meredith, 1978). Chest circumference varies in proportion to head circumference; on the average it is 1–2.5 cm (0.38–1 in) smaller than the head circumference.

The length of the gestation period is probably involved in the weight and proportion differences seen in newborns. Blacks have a 9-day shorter gestation period than Whites, which may account for their smaller size (Hardy, Mellits, 1977).

Body fat Weight-for-height and body fat measurements are not completely comparable, although they are both commonly used to judge the fat component. Weight-for-height measures are affected by bone density, musculature, and body fat, together or singly. When children of various races are weighed and measured, and their weight-for-height calculation is done, those in the more extreme percentile ranges should be assessed physically to see which components predominate. Children who are heavily muscled or who have robust skeletons should not be labeled obese.

The body fat component is easy to measure with skinfold calipers. The triceps, subscapular, and midaxillary skinfolds are considered reasonable estimates of the amount of body fat (Johnston, Hamill, Lemeshow, 1972). An even simpler method of determining body fat is visual inspection. Categorizing children by their degree of roundness, on a scale from 1 to 5 with 3 being normal, was found to correlate highly with the triceps measurement and with the Wetzel grid technique (Rauh, Schumsky, 1969).

Weight as well as height can be used to describe races, if caution is taken to determine which body component predominates. Asian and Black children are generally the lightest; White children are a little heavier, followed by American Indian, Eskimo, and Mexican-American children, who are the heaviest (Johnston, Hamill, Lemeshow, 1972; Johnston et al, 1978; Barr, Allen, Shinefield, 1972; Duncan, Smith, Briese, 1979; Heller, Scott, Hammes, 1967; Yanocheck-Owen, White, 1977).

Developmental Maturity Differences

Rates of growth and development also vary by race. Blacks tend to mature earlier than Whites. This may have resulted from the practice of breastfeeding a child for the first 3 to 5 years of life, when other appropriate foods were not available. In response to selective forces, children who matured faster and were capable of handling and eating adult foods at an early age survived better than slower maturers. In nomadic groups, such as some African tribes, children who learned to walk sooner were probably better off; not having to carry them around saved parental time and energy. The trouble with these speculations is that they cannot be confirmed. The only fact established is that growth and development rates differ by race.

Skeletal and neurologic maturity At birth Black children are more developmentally mature than White children. This advanced maturity is evidenced in both their skeletal and their neurologic systems (Falkner, Tanner, 1978). Blacks continue their lead until the age of 2 or 3 years in the neurologic system and until puberty in the skeletal system (Malina, 1970; Roche, Roberts, Hamill, 1978).

Children of the White and Oriental races start out similarly in skeletal maturity, but Oriental children start to lag behind White children by 3 to 6 months of age. They continue to lag until near puberty, when they begin a growth spurt more accelerated than that of the White children (Kimura, 1976; Roche, Roberts, Hamill, 1978). Even though their growth rate increases, they never reach the same final height as Black or White children. Neurologic maturity has not been studied enough to know whether Oriental children are similar, advanced, or retarded compared with White children.

The differences in skeletal maturity are due partially to genetic influences and partially to environmental influences: nutrition, disease, and altitude. Optimum environmental conditions advance skeletal maturity, while any conditions less than optimum retard it. A delay attributable to environmental stress does not necessarily decrease the individual's ultimate stature, because delayed maturity results in a longer period of growth. However, stress that is severe enough and occurs at a critical time during the growth period will stunt the final stature. The effects of mild stress, or even relatively severe stress, at a less critical growth phase are usually counteracted by compensatory growth.

Skeletal maturity is assessed from radiographs of the hand and wrist by comparing the degree of ossification of the bones with a standard. The two most common methods for assessing bone age were standardized on American and British White children (Weiner, Lourie, 1969). Consequently when bone ages are recorded, they are presented as advanced, retarded, or similar to the White standard.

Skeletal maturity estimates are primarily a research tool. They are not used in clinical practice unless pathology is suspected. Ordinarily, to assess development, a child's growth is plotted on a standard chart.

The recommended growth chart for children in the United States was developed from National Center for Health Statistics data (Hamill et al, 1979). It is adequate for children of all races. Adjustments have been worked out for racial, socioeconomic, and geographic variations. This information can be found in a series of publications by the United States Department of Health, *Vital and Health Statistics*, Series II.

The World Health Organization (WHO) also recommends use of the National Center for Health Statistics data and has developed its own growth charts based upon these data. The WHO charts allow for adjustments of the percentile ranges to fit different populations (World Health Organization, 1978, pp 1–35). See Figure 3.2 for an example of the WHO charts.

Clinicians must be cautious in using standardized growth charts on children of various racial groups. Growth charting is a screening, not a diagnostic, procedure. It indicates only how children compare with a standard. Furthermore, the standard makes no allowance for parental size, climate, or altitude. Hot climates and high altitudes have a stunting effect on

FIGURE 3.2 A WHO weight chart

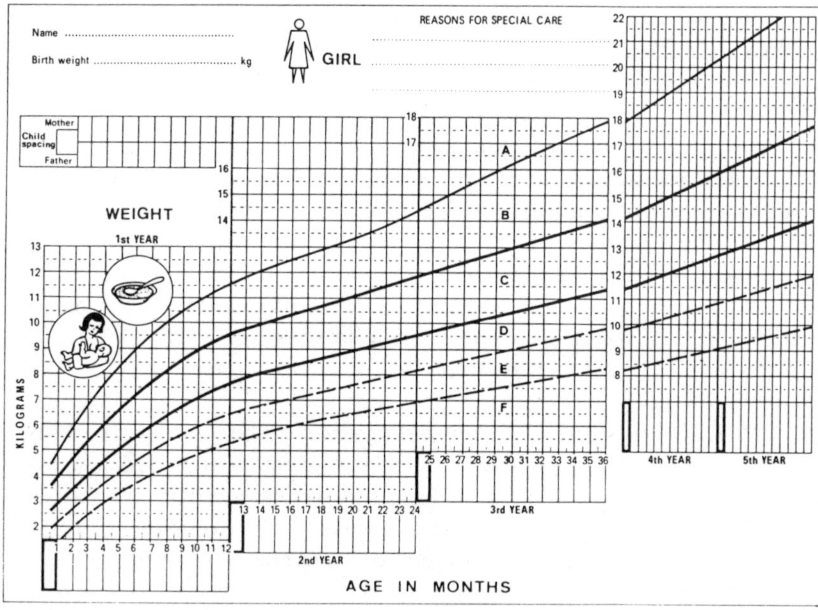

Source: World Health Organization, 1978, p 28.

Note: Channel A = greater than 97th percentile; channel B = between 97th and 50th percentile; channel C = between 50th and 3rd percentile; channel D = between 3rd percentile and –3 standard deviation; channel E = between –3 and –4 standard deviation; channel F = less than –4 deviation. The last two arbitrary but useful lines do not imply that specific numbers of the reference group lie below them; for this reason they are expressed as broken lines.

growth. These environmental influences are covered in detail later in this chapter.

If, after suitable allowance has been made, children still appear small for their age, have a low weight for their height, or show major changes in growth velocities by slipping down in percentiles, they need to be examined more thoroughly.

Tooth emergence Blacks are more advanced than Whites in the eruption of their permanent teeth. Black children's teeth erupt earlier, even when socioeconomic differences are taken into consideration. Poverty retards tooth emergence, but not by much (Garn et al, 1973), and poverty-level Black children are advanced in tooth emergence over middle-income White children. Early in permanent tooth eruption, Blacks are precocious compared to Whites; the White children catch up between 9 and 12 years of age (Freitas, Salazano, 1975). Japanese and Chinese children are slower than Black children in tooth emergence. White children are generally the last to erupt their permanent teeth.

Third molars start coming in by age 13 in most races; however, Whites are retarded compared with other races in this too. Rarely does a White child show emergence of a third molar before age 16 (Chagula, 1960).

Motor development Black infants are also ahead of White infants in motor development (Scott et al, 1955; Bayley, 1965). White children catch up by the time they enter school (Martinek, Cheffers, Zaichkowsky, 1978; Fuller, Friedrich, 1979).

This racial difference is probably partially environmental and partially genetic. Studies of Blacks of lower and higher socioeconomic status and studies comparing Blacks and Whites of average socioeconomic status show that the lower-status children are advanced over the higher-status children (Williams, Scott, 1953; Scott et al, 1955). In one study of 708 Black and White infants, the higher-status infants of both races were equal in motor development in the first 30 weeks of life; then the higher-status Blacks accelerated their motor development and became more similar to the lower-status Black infants (Scott et al, 1955).

One of the more commonly used screening tests of infant development, the Denver Developmental Screening Test (DDST), was standardized on 1,036 children, of whom 82% were White, 11% Spanish-surnamed, and 7% Black (Frankenburg, Dodds, 1967). The sample consisted of a higher proportion of upper-middle-class children than would be found in the United States population as a whole. Because of the small number of Blacks in the study, no attempt was made to check for racial differences. Health care providers should expect to see an accelerated pattern of development in

Black children when the DDST is used as a screening device. If a Black child lags much below the 50th percentile on motor development items, further diagnostic procedures are probably warranted.

Puberty Menarche has been studied in many groups around the world. Environmental and racial variables cause a wide discrepancy in the age of onset of menstruation. Menarche is delayed by altitude, poverty, and living in a rural setting (Eveleth, Tanner, 1976). Racial affiliation also seems to have an effect, but socioeconomic factors were not controlled in the studies that found wide differences. Menarche at around 12.5 years of age has been found in such diverse groups as middle-class girls in Santiago, Chile, upper-class girls in Istanbul, rural Naples girls, upper-class Hong Kong girls, and United States Black girls of all economic levels (Eveleth, Tanner, 1976). Menarche later than age 15 has been found in Bantu and Ruwanda girls in Africa (15–16.3 years), Mayan girls (15.1 years), Egyptian girls (15.2 years), and New Guinean girls (15.6–16.5 years) (Eveleth, Tanner, 1976).

A National Center for Health Statistics survey conducted between 1960 and 1970 indicates approximately a 3-month difference in menarche between Black and White girls in the United States, with Black girls starting at 12.52 years and Whites at 12.80 years (MacMahan, 1973). The racial difference holds when income and rural versus urban residence are controlled.

The contribution of race to the age of onset of menstruation is in some cases statistically significant but probably clinically irrelevant. Larger differences in menarche result from socioeconomic conditions. In the United States, clinicians will see few girls who have not achieved menarche by age 15; this age of onset is within the lower fifth percentile. Further examination is warranted for girls who have not reached menarche by 15 years.

The length of time available for growth is determined by the onset of puberty. This supposedly has a genetic component, but it is not well understood. What is known is that early maturers, those with early onset of puberty, are shorter than late maturers. Girls stop growing at around 16 years of age, about 3 years after menarche. Boys stop growing at around 18 or 19 years, which is also around 3 years after their onset of puberty. Men are generally taller than women, partially because of their longer growing period. The Y chromosome is thought to have an effect on the maturation rate. Evidence for this comes from studies on XYY individuals. It appears that the extra Y chromosome lengthens growth time. Individuals of XYY genotype are taller than expected from parental size (Cavalli-Sforza, Bodmer, 1971).

Determination of puberty in boys is less precise than in girls; consequently, few racial data are available. There is some suggestion in the literature that Black boys mature about half a year earlier than White boys.

Black girls also show similar precocity in the maturational states of puberty exclusive of menarche (Nankin et al, 1974; Foster et al, 1977).

Environmental Effects

Growth and development are influenced by the environment, which confounds the racial differences mentioned previously. The main environmental factors are climate, altitude, diet, and disease. These factors can be viewed as stressors, or in some cases enhancers, of growth when they occur at the extremes of what children generally experience. They are labeled stressors when they cause a slowing of growth and development or a decrease in final body size or response.

Climate Climate has an effect on the amount of body fat and the timing of tooth emergence. A study of American children living in Rio de Janeiro illustrates this (Eveleth, 1966). The children, from embassy and business families, belonged to the upper class, were of northern European extraction, and lived in American conditions in terms of food, health care, and socioeconomic status. They were studied twice a year over 4 years and were matched with four comparable groups of children in the United States. The tropically raised children weighed less than the temperate zone groups, and permanent tooth eruption occurred earlier in the tropically raised children.

In the United States, the finding that Black children are generally thinner than White children (Barr, Allen, Shinefield, 1972) may be partly due to the children's climate of residence as well as to the geographic region where each race evolved in the past. It is easy to see why a tendency toward thinness is advantageous in a hot climate (see chapter 2), but why teeth emerge earlier is not readily apparent.

Altitude Altitude affects birth weight in all races. Infants born at high altitudes weigh less than those born at sea level. Data on altitude effects come primarily from Peru and Colorado (Lichty et al, 1957; Frisancho, 1970; Haas et al, 1980). South American Indian and Colorado White infants born to mothers residing at altitudes around 10,000 feet are proportionately smaller than infants of these groups at lower altitudes. Their birth weights are 200–400 g (7–14 oz) less; they are also shorter and have smaller head circumferences. The smaller size of the infant is a response to the reduced amount of oxygen at high altitudes. The placental weight/birth weight ratio is around 12% higher, which increases the placental volume and surface area available for oxygenation of the fetus (Frisancho, 1970).

Children raised at high altitudes grow more slowly than children living in lower altitudes. Their growth period is also prolonged into their early 20s

(Beall et al, 1977; Palomino, Mueller, Schull, 1979). The slower, longer growth period at high altitudes enables these children to reach adult heights similar to those of racially matched controls at lower altitudes.

Diet and disease The environmental factors of diet and disease frequently act together. Where diseases are prevalent, diets are generally poorer. Less well nourished children are smaller and lighter than better nourished children. Because poorly nourished children are more susceptible to disease, the size discrepancy is aggravated. Harris lines, as mentioned earlier, correlate well with times of poor nutrition and disease.

Maternal nutrition also affects the weight of newborns. Studies of two diverse groups—poor, pregnant, New York City Black women and rural Guatemalan women—show that calorie supplementation is more effective at increasing birth weight than is protein supplementation (Lechtig et al, 1975; Rush, Stein, Susser, 1980). Protein supplementation in the New York City Blacks actually caused an excess of premature births and associated neonatal deaths, as well as a significant prenatal growth retardation up to 37 weeks of gestation (Rush, Stein, Susser, 1980). Dietary recall information on the New York and Guatemalan samples indicate that calorie intake was lower than optimum while protein intake was adequate. Evidently a sudden, large increase in dietary protein is not good for pregnant women.

Dietary supplementation programs in several areas of the United States, Canada, and South America show approximately a 40 g (1.33 oz) increase in average birth weight between treatment and control groups (Rush, Stein, Susser, 1980). However, the 40 g increase in birth weight does not sufficiently justify nutritional supplementation programs among chronically malnourished groups. Perhaps better justification for such programs could be found in maternal effects. Additional calories probably help pregnant women more than their fetuses. If the additional calories are continued during lactation, breast-fed infants would benefit. Lactating mothers might also gain energy needed to care for their families more adequately.

Nutritional requirements may differ by race, but the evidence is indirect and scanty. One challengeable nutritional requirement is the United States Recommended Dietary Allowance (USRDA) for calcium, which is probably inflated for all races. Lactose-intolerant populations—the majority of Blacks, Orientals, and American Indians and 10% of Whites—cannot ingest the recommended amount of calcium without consuming milk products equivalent to nearly a quart of milk per day, which causes gastrointestional discomfort in them. The recommended amount of calcium is probably inflated to match dietary practices in the United States and not based on solid evidence.

Surface Features

A few surface features pertinent during the growth period will be mentioned here, particularly orofacial characteristics, earwax, birthmarks, and sweat chlorides.

Orofacial characteristics

Lip pits and *preauricular pits* were mentioned in chapter 2. Their presence is generally detected in childhood. They are normal physiologic variations that are rather inconspicuous and cause no problems.

Another oral characteristic mentioned in chapter 2, *cleft uvula*, is also detected in early childhood. It is also inconspicuous and causes no problems, but there is some evidence that it is a subclinical manifestation of cleft lip and palate. The connection with cleft lip and palate is not strong enough to be of much use in genetic counseling, but health care professionals can indicate to parents that the child was lucky not to have a more serious cleft.

Variation in the timing of permanent tooth eruption was mentioned earlier in this chapter. Another tooth variation is congenitally missing third molars (agenesis). Blacks and Australian Aborigines are the least affected by third molar agenesis; only 1–2% of them have it. Chinese, Japanese, and Eskimos are the most affected, around 30%. Whites and Native Americans are intermediate, approximately 10–15% (Brothwell, 1963). Third molar agenesis has no particular health significance; it is merely an interesting difference between groups.

Racial variation also occurs in *neonatal teeth*. It is rare to see a White baby born with teeth (1 in 3,000), but among some American Indian groups, it is not uncommon (approximately 1 in 11 in Tlingit Indians) and approximately 1 or 2 in 100 Canadian Eskimo babies have neonatal teeth (Jarvis, Gorlin, 1972). Generally, the presence of neonatal teeth does not interfere with breast-feeding.

Leukoedema, a benign oral lesion mentioned in chapter 2, has been considered a condition of adult mouths, but it may start as early as 3 years of age. Blacks have the condition most commonly; one investigation to determine its incidence in Black children found that 51% between ages 1 and 18 had leukoedema (Martin, Crump, 1972). The lesions increased in severity with age and were related to the degree of skin pigmentation: darker-skinned children had more severe leukoedematous lesions.

Earwax

Chapter 2 contained information on earwax (cerumen) that is pertinent for care of children. There are two types of earwax, and practitioners using an otoscope need to recognize the differences. Wet wax, the most common type in Whites and Blacks, is dark brown and moist. Dry wax, most common in Orientals and Native Americans, is gray and dry. It is flaky and may form a thin mass that lies in the ear canal. The Japanese call this type of wax rice-bran earwax, which gives a good picture of what it looks like (Matsunaga, 1962). The Japanese also have descriptive names for wet earwax: cat earwax, oily earwax, and honey earwax. Clinicians not aware of the two earwax types might confuse the flaky appearance of the dry wax with a dry, eczematous lesion.

Sweat chlorides One piece of information on sweat chlorides needs to be added to that covered in chapter 2: the connection between sweat chlorides and cystic fibrosis. Children with cystic fibrosis have higher sweat chloride levels than normal children and consequently a generally lower galvanic skin resistance (Batson, Young, Shepard, 1962). Black children have lower sweat chloride levels than White children, and this may be related to the rarity of cystic fibrosis in Black children.

Birthmarks The most common birthmarks that show racial variation are mongolian spots, mentioned in chapter 2, which are more common in Black, Oriental, American Indian, and Mexican-American babies. Another birthmark category is pigmented marks (Jacobs, Walton, 1976). These marks, characterized as sharply demarcated macules, light tan to dark brown in color depending on the child's skin color, are most common in Black children (20%) and less common in Whites, American Indians, and Mexican-Americans (2–3%).

Common Clinical Measurements

Surprising as it may seem, some routine clinical measurements are subject to racial variation. Blood pressure, heart rate, and body temperature differ by race, as do the laboratory measures of hemoglobin/hematocrit, serum cholesterol, serum transferrin, and two amniotic fluid constituents. Apgar scores also differ between populations.

Blood pressure Black adults have elevated blood pressure more often than White adults, but the difference is not apparent in children or adolescents (Reed, 1981; Rowland, Roberts, 1982). In fact, the rise in mean blood pressure for Blacks, compared with Whites, does not start until after age 35 (Rowland, Roberts, 1982).

Evidence indicates that, whatever the race, overweight and obese children have higher blood pressure than normal to underweight children (Walker et al, 1978; Lynds, Seyler, Morgan, 1980). A study of 1,692 Black children from 5 to 11 years of age showed that 195 children had elevated systolic pressure and 199 had elevated diastolic pressure (Lynds, Seyler, Morgan, 1980). The children with elevated systolic pressure were three times as likely to be obese as the other children. Diastolic elevations showed a similar but less intense correlation with obesity.

Pulse and temperature Newborn Black infants have significantly higher heart rates during sleep than White newborns, even when socioeconomic level is controlled (Schachter, Lachin, Wimberly, 1976). Although socioeconomic level affects heart rate, it does not account for as much variance in the rate as does race. The difference is 6 beats per minute faster for Black infants when socioeconomic level is controlled; when the socioeconomic level is not controlled, the difference rises to 9 beats per minute. A difference of 6 to 9 beats per minute is large enough to be clinically significant. To date the reason for the racial difference is unknown, but there might be a relationship between this heart rate difference and the later blood pressure difference between the races. Another area that needs further research is whether this difference in pulse rate continues into childhood and adulthood.

Newborn body temperature may also vary by race, but the evidence is sparse and contradictory. Two studies of Black and White infants give opposite findings. One study of 116 newborns (half Black, the other half White), using axillary temperature measurements, indicated a statistically significant 0.5° higher body temperature for the White newborns (Whitner, Thompson, 1970). The other study, of 247 newborns (approximately half White and half Black), using rectal temperature measurements, found no difference (Schachter, Lachin, Wimberly, 1976). It could be that the core temperatures of these two groups are the same, but the lighter and thinner Black newborns have a lower surface temperature. Whatever the explanation, the difference is not clinically relevant, but it would be interesting to know if a temperature difference continues into childhood and adulthood.

Laboratory measures Many data have been accumulated on *hemoglobin/hematocrit* differences between the races, yet a good deal of information is still missing. It can be stated with reasonable certainty that the normal hemoglobin level for Blacks is 1 g lower than for other groups (Garn, Smith, Clark, 1975; Johnson, Abraham, 1979). The data also indicate that, given similar socioeconomic status, Native Americans, Mexican-Americans, Oriental Americans, and White Americans do not differ (Rode, Shephard, 1973; Heald et al, 1974; Ershoff, 1975; Johnston et al, 1978).

Socioeconomic status has a strong effect on hemoglobin/hematocrit levels: the poor have lower levels than the middle and upper classes (Heald et al, 1974). The lower blood levels in Blacks remain even when socioeconomic status is controlled, however (Garn, Smith, Clark, 1975). There is some indication that, given similar socioeconomic conditions, Asian Americans and Mexican-Americans may have higher levels than Whites (Ershoff, 1975; Centers for Disease Control, 1980). The clinician should consider an individual's racial group before instituting iron therapy. Overvigorous application of normal White values to Blacks could cause unnecessary constipation.

Serum cholesterol levels are also affected by socioeconomic status. Children from higher socioeconomic levels tend to have higher cholesterol levels (Levy et al, 1976). But racial affiliation has an even stronger effect. Black children have higher serum cholesterol levels than White children, despite the lower socioeconomic level of the Blacks in the sample (Levy et al, 1976). In adults, the serum cholesterol differences are reversed; adult Blacks have lower serum cholesterol levels than White adults (Fulwood, Abraham, Johnson, 1980). At birth the two races have similar cholesterol levels (Glueck et al, 1977). Why the values start differing in childhood, and then reverse direction in adulthood, is unknown.

Data on Pima Indians show that they have considerably lower cholesterol levels than Whites (Savage et al, 1976): 20–30 mg/100 mL lower in childhood, increasing to a 50–60 mg/100 mL difference in adulthood. The Black-White differences are much less, only 5 mg/100 mL, Blacks being higher (Fulwood, Abraham, Johnson, 1980). The effect of diet was not controlled in either of these studies.

Serum albumin and *serum transferrin* were measured in Black and White children, 1 to 3½ years of age (Roode et al, 1975). Albumin did not differ by race, but transferrin did. The mean values for serum transferrin ranged from 200 to 400 mg/100 mL, with the White children having a mean value of 319.7 and the Blacks a mean of 341.4. This difference was mainly due to a considerable difference between the girls (340.4 for Blacks, 304.4 for Whites). The boys differed, but not as much (342.9 for Blacks, 337.7 for Whites). The higher serum transferrin levels in Black children may be due to their lower hemoglobin/hematocrit levels (Ritchie, 1979). Transferrin levels increase in the presence of anemia. If the hemoglobin/hematocrit levels are normally lower in Blacks, then higher tranferrin levels should be considered normal.

A laboratory measure that indicates fetal pulmonary maturation is the *lecithin/sphingomyelin ratio* in amniotic fluid. The ratio is used to calculate the risk of respiratory distress syndrome in premature infants. This ratio differs between Blacks and Whites, as does the pulmonary maturity it predicts (Olowe, Akinkugbe, 1978). Blacks have higher ratios than Whites from 23 to 42 weeks of gestation. Lung maturity, measured by a lecithin/sphingomyelin ratio of 2.0, is reached one week earlier in Blacks than Whites (34 versus 35 weeks). The risk of respiratory distress syndrome is 40–50% for a ratio score between 1.5 and 1.9 for White prematures but not for Blacks. Black premature infants have a much lower risk of respiratory distress syndrome at the same low ratio scores. The specific risk for Blacks has not been calculated, but the disease is 40% less likely to produce death in premature Blacks than Whites (Farrell, Wood, 1976). When the lecithin/sphingomyelin ratio is determined before induction of labor or elective cesarean section, the racial difference should be considered in making the decision.

Apgar scores also differ somewhat between Blacks and Whites. Black babies have lower 1 minute Apgar scores than do White babies; 88.4% of Black babies reach 7 or higher, compared with 91.3% of White babies (Querec, 1981). By 5 minutes, the difference is reduced to 1.8% (96.4% for Blacks, 98.2% for Whites). These differences are not clinically significant, but they point out the variation that can occur in infant condition because of socioeconomic levels, specifically the lowering of Apgar scores due to low birth weight and the mother's young age, low education level, few prenatal visits, and unmarried status. These conditions are all more prevalent in Black than White mothers (Querec, 1981).

Disease Differential

Children of various races react differently to certain diseases. In some cases a disease is more or less severe in a specific group; in other cases there is greater susceptibility to a disease. Chapter 4 is devoted to disease, but certain conditions that affect children are mentioned here: neonatal jaundice, ABO incompatibility, sickle cell disease, otitis media, and the incidence of congenital abnormalities.

Neonatal jaundice

Clinical jaundice is present in over 50% of Oriental and American Indian babies (Horiguchi, Bauer, 1975; Saland, McNamara, Cohen, 1974). Bilirubin levels are correspondingly high in these groups compared with Black and White infants. Approximately 25–40% of Oriental and American Indian infants have bilirubin values above 12 mg/100 mL, the value considered the upper range of normal (Horiguchi, Bauer, 1975; Postl, Nelson, Carson, 1982); 15–23% have values above 15 mg/100 mL, which is considered pathologic for White infants (Horiguchi, Bauer, 1975; Fisher et al, 1978). Norms should be set for Orientals and American Indians. Not only do they have higher normal bilirubin values, but also the blood level peaks later, on the fifth or sixth day compared with the second to third day for other races (Saland, McNamara, Cohen, 1974; Horiguchi, Bauer, 1975). Breast-feeding enhances the bilirubin values in all babies, but the effect is stronger in the Oriental and American Indian babies (Saland, McNamara,

Cohen, 1974). The higher range of bilirubin levels in Oriental and American Indian babies does not appear to be associated with a higher incidence of kernicterus (Saland, McNamara, Cohen, 1974).

ABO incompatibility A more serious type of neonatal jaundice showing racial variation is ABO incompatibility. It can lead to early fetal loss (spontaneous abortion) when the fetus is A or B blood type and the mother is O blood type. Fetal wastage caused by ABO incompatibility occurs in around 18% of maternal-fetal ABO incompatible pregnancies (Lauritsen, Grunnet, Jensen, 1975). At term, about 1% of the surviving infants are affected with ABO hemolytic disease. Here is where the racial differences become apparent. Black infants are over twice as likely as White infants to manifest the disease (Bucher et al, 1976; Kirkman, 1977).

Because neonatal jaundice is harder to detect in Blacks, and because Black infants have a higher risk of developing ABO hemolytic disease, it is generally recommended that all mothers and infants be ABO blood-typed at delivery. Infants from incompatible pregnancies should have Coombs tests and bilirubin levels measured before hospital discharge, which should be delayed until the bilirubin level has started to diminish (Diamond, 1976; Peevy, Wiseman, 1978). Phototherapy is usually sufficient to reduce the bilirubin level (Diamond, 1976; Peevy, Wiseman, 1978).

Sickle cell disease Sickle cell disease is covered in greater detail in chapter 4, but, since most patients with the disease are children, data on population differences will be mentioned here. The disease is commonly considered a condition affecting Blacks exclusively, but that is not really so. Although Blacks have a much higher incidence of sickle cell disease than other races, occasional cases have been seen in children of Greek, Italian, Turkish, and Arabian ancestry (Campbell, Oski, 1977) and individuals of Germanic, Italian, and Native American origin (McGrew, 1973).

There seem to be two levels of severity in sickle cell disease—a less severe form, present more often in the non-Black population, and a more severe classic form most prevalent in Blacks (Campbell, Oski, 1977; Dover, Boyer, Pembrey, 1981). See the discussion in chapter 4.

Otitis media Otitis media is much more prevalent in American Indians, Alaskan Eskimos, New Zealand Maori, and Australian Aborigines than in either Blacks or Whites (Brody, Overfield, McAlister, 1965; Johnson, 1967; Manning, Avery, 1974). When United States Blacks and Whites are compared, Whites have a higher incidence than Blacks (Griffith, 1979). There is much speculation on these differences, but few hard data. Socioeconomic conditions, specifically poor nutrition, poor hygiene, and crowding, are cited as possible risk enhancers (Johnson, 1967; Reed, Struve, Maynard, 1967). The lack of breast-feeding has been suggested as a precipitator (Manning, Avery, 1974). Anatomic differences in the size and shape of the eustachian tube have also been implicated (Timmermans, Gerson, 1980). Since otitis media was not prevalent during the early childhood of most adult Eskimos (Manning, Avery, 1974) and Orientals do not seem to get otitis media in extraordinary proportions, an environmental cause, either socioeconomic, nutritional, or climatic, is probably the culprit. Because recurrent otitis media leads to hearing loss, health workers should institute vigorous treatment and counseling to high-risk groups.

Congenital abnormalities Numerous congenital abnormalities show differences by racial group, but only a few studies have compared specific races. Table 3.1 was derived from these and some cross-study comparisons. It shows that Blacks have a very high incidence of polydactyly and supernumerary nipples and a moderately high incidence of cleft gum or uvula. They have a low incidence of the other conditions tabulated. Whites, in contrast,

TABLE 3.1 Incidence rates of selected congenital abnormalities

Congenital abnormality	Incidence per 1,000 population: Black	White	Chinese	Japanese	American Indian	Hawaiian
Supernumerary nipples	11.40	0.90	—	—	—	—
Polydactyly	10.97–12.80	1.16–1.40	0.69–1.53	—	—	—
Syndactyly	0.59	1.77	0.23–0.27	—	—	—
Hypospadius	1.21–1.95	2.39–3.05	0.60	—	—	—
Cleft gum or uvula	3.77	1.46	—	—	—	—
Cleft lip and palate	0.24–0.62	1.14–1.34	0.22–0.43	2.13	1.38	—
Talipes equinovarus	1.24	1.12–2.32	0.09–0.39	0.54	—	6.8

Sources: Derived from Christianson et al, 1981; Erickson, 1976; Emanuel et al, 1972; Ching, Ching, Nemechek, 1969; Niswander, Adams, 1967; Chung, Myrianthopoulos, 1968.

have a low incidence of polydactyly, supernumerary nipples, and cleft gum or uvula, but a moderate to high incidence of the other conditions. Chinese have a low incidence of syndactyly, hypospadias, and talipes equinovarus (clubfoot) and a moderate incidence of the other conditions. Japanese have a high incidence of cleft lip and palate and a moderate incidence of talipes equinovarus. American Indians have a moderate incidence of cleft lip and palate. Finally, Hawaiians have a very high incidence of talipes equinovarus.

Congenital anomalies of the nervous system also differ by race; these are discussed in chapter 6. In fact, those interested and patient enough can tease out racial and population differences in many other congenital abnormalities.

Other Variations

The last two variations among groups in growth and development discussed in this chapter occur in lactose intolerance and twinning rates.

Lactose intolerance Lactose intolerance varies by race. Generally the vast majority (80–90%) of the populations of all races, except Whites, are lactose intolerant in adulthood. Even in the White race, 10–15% are lactose intolerant. This condition is covered in greater detail in the section on enzymatic and biochemical diversity in chapter 4, but health professionals who work with children can use some information on inheritance, age of onset, and nutritional adjustments. Unfortunately, scientific data on these are very scarce.

Though lactose intolerance is certainly familial, whether it is dominant or recessive or possibly polygenic is still unclear (Newcomer et al, 1977; Johnson, Cole, Abua, 1981).

The age of onset is no better described. There is fairly good evidence that it does not appear much before 5 years of age (Lebenthal, Antonowicz, Schwochman, 1975; Newcomer et al, 1977), but a variable onset has been noted, up to the teens or early 20s (Caskey et al, 1977).

Nutritional advice for lactose-intolerant individuals is controversial. No hard data are available to demonstrate that the USRDA calcium requirements are necessary for all races, let alone the White Americans for whom they were devised. It is safe to suggest that children who suffer gastrointestinal symptoms from the ingestion of milk not be required to drink it. Clinicians should remember that the bones of Black children are denser than those of White children, and their teeth are freer of decay, though they consume less milk (Villa Dresser, Carroll, Abraham, 1978). Meeting the USRDA calcium requirement without using milk or milk products is difficult, unless calcium supplements are used. If a family's traditional diet includes foods that contain calcium, as many vegetables do, and the children appear well nourished, it may be conservative practice to avoid urging nutritional changes to meet standard calcium requirements.

Twinning rates Monozygotic twinning is similar in all races, but dizygotic twinning differs, being highest in Blacks (up to 5% of births). The rate is about 2% in Whites, and 0.5% in Asians (Bulmer, 1970). In the United States a multiple-birth ratio is calculated yearly; it is the number of live births in multiple deliveries per 1,000 live births. This ratio runs around 18.6 for Whites, compared with 24.2 for Blacks. The ratio also goes up with age; for instance, mothers 35 to 39 years of age have a ratio of 27.1 compared with 13.1 for mothers 15 to 19 years of age (National Center for Health Statistics, 1981). Blacks and Whites are affected similarly by the age change.

Summary

The existence of population differences in rates of growth, in body size and shape, in specific features, even in some of the common health measurements, and the fact that diseases differentially affect children of different populations should convince nurses and physicians to maintain an open attitude about health measurements. In the absence of specific indicators to the contrary, a physically active and progressing child who may be a little thin or short by White standards, but is alert and without obvious signs of illness, is probably healthy; clinicians should not be too quick to suspect growth retardation or malnutrition. They need to question the norms applied to each race and to document the full range of normality for all children.

AGING DIFFERENCES

Contrary to popular opinion, populations do not seem to differ in attainment of extreme old age. The so-called long-lived groups—Russians from the Caucasus mountains, Hunzas from the Karakoram mountains in West Pakistan, and Ecuadorans from the Andes mountains—do not live to be any older than the general population in the United States. When these reputed centenarians were investigated, they were usually found to be in their 70s or 80s (Leaf, 1982). Rarely was an individual found who had reached 100 years of age. Some of the reasons for age exaggeration were the desire to advertise the health benefits of living in the Russian mountains, a practice of taking on the identity of a parent to avoid military conscription, the desire to be a celebrity, and the increased prestige of aged members of the group.

The upper end of the life span appears to be similar in all populations. For instance, the 1970 US Social Security Administration figures on individuals receiving benefits revealed 6,200 people 100 years of age or older in the United States. These figures are comparable with Japanese birth and death records that have been maintained for the past 100 years (Leaf, 1982). However, the US census figures for the same year (1970) showed 107,467 centenarians, 17 times more than the social security figures. This demonstrates that Americans exaggerate their ages too, but they have not received as much publicity as more exotic mountain folk.

While the extreme end of the life span does not differ between groups, life expectancy does. Socioeconomic and health conditions affect life expectancy. Consequently, in the United States, Whites live longer than "non-Whites." Sex also has an effect: women live longer than men.

Unfortunately the US data on life expectancy separate White and "non-White" categories only. This does not allow an adequate picture of the several racial groups involved. Non-White includes, among others, Blacks, Orientals, Native Americans, and East Indians. The data show that White women have the highest life expectancy at 78.1 years, followed by non-White women at 74.0 years, White men at 70.5 years, and non-White men at 65.3 years (National Center for Health Statistics, 1981). Blacks, American Indians, and Eskimos have lower life expectancies than Asian Americans and East Indians (Metropolitan Life Insurance Company, 1977). Somewhere around 10% of the non-Whites have life expectancies similar to Whites; this inflates the non-White figures to some extent.

There are signs of improvement in life expectancy for non-Whites. Vital statistics records indicate that the non-White population shows a more pronounced increase in longevity than Whites. Between 1970 and 1978, non-Whites added 4.2 years to their life expectancy, while Whites added only 2.4 years (Metropolitan Life Insurance Company, 1980). Non-White females have improved their life span faster than non-White males during the same period—4.5 years for females, 4.0 years for males—so the sex difference in longevity has widened.

Vital statistics data also show a switch in longevity patterns around the age of 65–70, when non-White survival becomes better than Whites' (Metropolitan Life Insurance Company, 1980). This probably occurs because older non-Whites are very healthy individuals who have survived less ideal living conditions than their White counterparts.

Increases in life span are not expected to go on without end. Gerontologists consider 85–95 years to be the normal life span (Fries, 1980; Hayflick,

1981). Elimination or long postponement of the sequelae of cardiovascular diseases and cancers is calculated to improve the present average life span by only about 14 years (Hayflick, 1981).

The aim of medical science is the "compression of mortality" into the final years of life, according to one gerontologist (Fries, 1980, p 130). A more positive view is that science will delay the usual chronic diseases of old age until after natural death has occurred. One pathologist has suggested that physicians should view the cause of death in healthy oldsters as senescence rather than a specific disease (Kohn, 1982). He views aging as "a universal, progressive decline in physiological function to the point where life cannot be maintained in the face of otherwise trivial tissue injury" (p 2793). It seems less ghoulish to consider death in old age the normal end to life, rather than pathologic.

The normal physiologic accompaniments of aging are discussed in the remaining sections of this chapter, in terms of race and sex differences. Body size, shape, and composition, with the attendant surface manifestations, are covered first. Then developmental differences such as menopause and sensory changes are described, and the adjustments clinicians must make when taking common health measurements are specified. A short section on disease susceptibility precedes the final section on prediction and promotion of longevity.

Body Size, Shape, and Composition

The major changes of old age are loss of height, redistribution and recomposition of weight, and continued growth of the skull and cartilage.

Loss of height and bone density Human beings lose height as they age because the intervertebral discs become thinner and the height of the vertebral bodies decreases (Friedlaender et al, 1977; Ericksen, 1978a, 1978b). Not much height is lost this way—only about 0.6 cm (0.25 in) per decade starting in the fifth decade of life (Friedlander et al, 1977; Trotter, Gleser, 1951).

Measuring the loss of height with age is complicated by a secular trend toward increased height. Long-term longitudinal studies are difficult; scientists age and lose interest in such studies as fast as subjects do. One of the few longitudinal studies of old men started with a cohort of 170 subjects; 16 years later, only 22 men were available for study (Parizkova, Eiselt, 1971). This study verified what the cross-sectional studies, which use statistical manipulations, have indicated: height does decrease with age. In 16 years the men lost, on the average, 1.5 cm (0.63 in) in height, most of it through a decrease in their sitting height, which is a measure of the length of the spine. The height decline appears to be similar for both Blacks and Whites of both sexes (Trotter, Gleser, 1951; Abraham, Johnson, Najjar, 1976). No data are available for other races.

Older persons also lose bone density. This loss, or osteoporosis, occurs in the cortical areas of long bones and is partially compensated by increases in bone in the periosteal area. Consequently, long bones increase in diameter as they increase their medullary space. A physics principle helps explain this phenomenon: as the radius of a cylinder increases, the rigidity of the structure increases rapidly (Pfeiffer, 1980). This process helps maintain bone integrity as bony substance lessens. A problem may occur because of the increased diameter of the medullary cavity; for example, an individual with an old hip prosthesis may experience a loosening of the artificial hip at its attachment to the femur (Ericksen, 1979).

Extensive bone loss can lead to fractures in older individuals. Bone loss is a normal process in aging, but individuals who start out with light bones or who incur greater than normal bone loss as they age reach a point of bone fragility where fractures occur from little or no trauma. Osteoporosis of this degree is pathologic. The extent of normal osteoporosis varies by race and sex. Blacks have denser bones than Whites; Whites have denser bones than Chinese, Japanese, Eskimos, and some Native Americans (Trotter, Peterson, 1966; Garn, Pao, Rihl, 1964; Mazess, Mather, 1975; Ericksen, 1976). Males have denser bones than females (Trotter, Peterson, 1966; Garn, Pao, Rihl, 1964; Mazess, Mather, 1975; Ericksen, 1976). Consequently women more often suffer fractures from extensive osteoporosis than do men; but no data indicate that Orientals and American Indians have more fractures in old age than Whites do (Mazess, Mather, 1975). The shorter, stockier body build of Eskimos, with their tendency to be big boned in relation to body size, has been suggested as the reason for their dearth of osteoporotic fractures (Mazess, Mather, 1975). Perhaps a similar explanation might hold for other Oriental and Native American groups.

Women lose bony substance differently from men. Men reach their maximum bone density in early adulthood, and then slowly decline. Women reach their maximum bone density about 10 years later than men, maintain it until around age 50, and then experience a more rapid decline than men do (Morgan et al, 1967; Odland, Mason, Alexeff, 1972).

Weight changes Body composition also changes with age. People lose muscle but gain fat (Winterer et al, 1976; Forbes, 1976; Borkan, Norris, 1977; Parizkova, Eiselt, 1980). The loss of muscle results in thinner extremity circumferences and increased abdominal girth (Damon et al, 1972; Parizkova, Eiselt, 1980).

Evaluating weight change with age is difficult because of the simultaneous muscle loss and fat gain, and because of a secular trend toward increased height over the past 50 years or so (Abraham, Johnson, Najjar,

1977). Both males and females have participated in this secular trend. Extrapolations from cross-sectional data indicate that women gain weight rapidly until approximately age 45 and then gain slightly until about 70, when they begin to decline in weight. Men gain weight rapidly until about 30, gain slightly until around 50, and then begin a decline (Abraham, Johnson, Najjar, 1977). Men's weights follow the gain and loss of muscle mass more closely than do women's weights. Women put on most of their weight as fat.

Skull and facial changes Head and face changes are similar in men and women: the skull bones and cartilage continue to grow with age; the jawbone decreases in height but increases in width, and the earlobes and nose grow longer (Israel, 1977; Overfield, Call, 1983; Ham, 1974). These changes produce the distinctive head of the oldster, which looks large in proportion to the rest of the body.

Besides these changes, many aged individuals lose their teeth: 51% of individuals 65 years of age or older are edentulous (Burnham, 1974). This brings the nose and chin into closer proximity. The better dental care received by younger generations should reduce the number of edentulous old people in the future.

The head and face changes occur similarly in all races, but tooth loss differs by race. Blacks of all ages have less tooth decay but more periodontal disease than Whites. Despite this, only 43% of Blacks 65 or older are edentulous, compared with 51% of Whites (Burnham, 1974).

Surface Manifestations

Wrinkles and gray hair are usually the layperson's first clues to aging. These are also good clues for health workers. A study conducted on over 400 Japanese demonstrated that gray hair correlates well (r = .72) with age (Hollingsworth, Hashizume, Jablon, 1965). A study of a very large sample (3,875) of Australian Whites confirmed this finding and provided the additional information that no significant differences in hair graying occurred between the sexes (Keogh, Walsh, 1965).

Obvious racial differences occur in the rate of hair graying, though. Whites gray faster than the other races. Specific studies to demonstrate this differential rate are not available, but the difference in graying rate between fair- and dark-haired individuals of unspecified race has been studied (Burch, Murray, Jackson, 1971). The finding was that 66% of fair-haired individuals are fully white by age 60, while only 37% of dark-haired persons are totally white by the same age.

Wrinkles also correlate well (r = .60) with age (Hollingsworth, Hashizume, Jablon, 1965) and differ between races. Blacks, Orientals, American Indians, and Eskimos appear to wrinkle later than Whites. Others have mentioned this, and I have observed it, but empirical support is lacking (Brues, 1977; Szabo, 1975). Health care workers may over- or underestimate age when they deal with clients from a race different from their own. Whites tend to underestimate the age of Black, Oriental, and American Indian clients, while Black, Oriental, and American Indian health workers tend to overestimate the age of White clients.

Dry skin is one of the more annoying surface manifestations of aging; it is inevitable in individuals over 70 years of age (Kligman, 1979). The cause of dry skin in the aged is unknown. It is not due to water loss or lack of sebaceous gland activity (Kligman, 1979; Pochi, Strauss, Downing, 1979). Sebaceous gland secretions do decline with age, but not until around 80 years in men. Women experience a slow decrease in sebaceous gland activity between 50 and 70 years, after which the activity levels off. Daily bathing and using emollients seem to help, whatever the cause (Gilchrest, 1982).

Developmental Changes

Some body functions and measurements decrease with age; others increase. As already noted, height and muscle mass both decrease, the jawbone shortens in height but increases in width, and both the skull and the abdomen increase in diameter.

Menopause

In women the reproductive function disappears altogether. Menopause occurs at a mean age of approximately 50 years in most populations (Brand, Lehert, 1978; Treloar, 1974; Admundsen, Diers, 1973; Frere, 1971). There appears to be no racial difference in mean menopausal age, but there is a socioeconomic difference. Poor women experience an earlier menopause than do better-off women (MacMahon, Worcester, 1966). Fertility begins to decline at least 15 years before menopause. The decline is slight but statistically significant in the early 30s; it is marked after 35 years of age (CECOS Federation, Schwartz, Mayaux, 1982).

Sensory changes

One of the earlier signs of aging is the need for reading glasses or bifocals. For women this occurs in the early 40s, for men in the late 40s to early 50s (Roberts, Cohrssen, 1968). Before 10 years of age, people begin to lose focusing power. By age 30, accommodative power has halved, and by 50 it is down to 1–2 diopters (Tripathi, Tripathi, 1983). Near vision is seriously affected at around 3.5–4 diopters.

Both near and far vision decline with age, but near vision declines earlier; few people over 55 can see close objects clearly without correction (Roberts, 1964). Far vision declines more slowly; by 65 years of age, 10% of individuals still have 20/20 vision.

Dim light that does not compromise the seeing ability of a young person does give an older individual trouble. Glare and intensely bright light are also bothersome (Hatton, 1977). Lighting conditions in hospitals and waiting and examining rooms should be adjusted accordingly. Hospitalized individuals who wear glasses should have their glasses kept handy, protected, and clean.

Hearing does not decline as early or as much as vision. In a national survey, only 8% of individuals over 65 years of age admitted to severe hearing problems (Gentile, 1975). A more objective measure, that of speech reception, indicates that the ability to hear speech sounds at 20–30 decibels declined to 82% of those tested at 55 to 64 years, and to 67% of those 65 to 74 years of age (Rowland, 1980). There do not appear to be major differences in hearing decline between the sexes, or between Blacks and Whites (the only two races measured).

The sense of taste slowly declines for salt and quinine sulfate, a bitter substance (Grzegorczyk, Jones, Mistretta, 1979; Weiffenbach, Baum, Burghauser, 1982). Stuides on the ability to taste sucrose are inconsistent: in one, the ability remained the same (Weiffenbach, Baum, Burghauser, 1982); in another it decreased slightly (Moore, Nielsen, Mistretta, 1982). The taste for citric acid, a sour substance, remains the same with age but does show a difference by sex: women can taste lower concentrations of it than men (Weiffenbach, Baum, Burghauser, 1982). These taste responses point out that there is not an overall decline in taste sensitivity. Where taste does decline, the degree of decline is slight. Decreased food enjoyment by some older persons is probably due to several factors. "Taste is only one component of flavor perception, which includes gustatory, olfactory, tactile, temperature, visual, and possibly even auditory components" (Moore, Nielsen, Mistretta, 1982, p 68).

The sense of touch changes little if at all in older individuals (Thornbury, Mistretta, 1981; Potvin et al, 1980). Pain sensation declines somewhat (Schludermann, Zubek, 1962; Sherman, Robillard, 1964). The ability to feel vibrations decreases considerably (Potvin et al, 1980; Verrillo, 1980). None of these sensations has much effect on normal functioning.

Changes in laboratory values Reference values for many common laboratory measurements have not been established for the elderly. Normal ranges are generally fixed by studying healthy young adults and children, or blood donors. Few studies have been done using healthy old people. One exception is a study of 73 people between the ages of 74 and 94, applying 53 different tests (Jernigan et al, 1980). All in all, 15% of those tested had an abnormal finding in one or more of the tests. Many tested above normal in urea, creatinine, glucose, triglycerides, immunoglobulins M and A, and α_1-

antitrypsin. Many were below normal in phosphorus, total protein, amylase, and osmolarity tests.

Tests for iron deficiency are among the better studied laboratory measurements in older age groups. It has been established that healthy old persons have hemoglobin and hematocrit values within the range of younger persons (Devine, 1967; Jernigan et al, 1980; Garry, Goodwin, Hunt, 1983). The sex difference in these blood values (males have higher readings than females) decreases with aging (Devine, 1967). Women experience a rise in these measures up to the age of 60, and then a slight decline, while men show a constant decline from the age of 30 to the end of life. By age 60, men and women have quite similar blood measures. The racial difference in hemoglobin levels continues into old age. Blacks have lower hemoglobin levels than Whites, but they follow the same sex pattern of increasingly similar hemoglobin levels with age (Devine, 1967).

Another laboratory measurement that is fairly well documented over the life span is blood glucose. Whether measured by the glucose tolerance test or by glycosylated hemoglobin determination, the blood glucose level rises with age (Gordon, 1964b; Arnetz, Anders, Tores, 1982). The mean blood glucose level, after a glucose challenge, rises from 99.7 mg/100 mL at 18–24 years of age to 166.3 mg/100 mL at ages 75–79 (Garst, 1966). Mean glycosylated hemoglobin determinations are 6.9% at ages 50–69 and rise to 8.62% at ages 80–89 (Arnetz, Anders, Tores, 1982).

Women have higher blood levels than men. This sex difference widens with age, being 9.5 mg/100 mL at ages 18–24 and 24.1 mg/100 mL at 75–79 years (Garst, 1966). There is also a racial difference: Blacks have slightly higher levels than Whites, but the difference is not apparent between individuals of similar socioeconomic status.

The increase in blood glucose levels with age and the higher levels in women indicate that the glucose levels used to diagnose diabetes may need to be adjusted upward for these categories. Or the data may be taken at face value; that is, women have diabetes more than men, and the incidence increases with age. The issue needs further study, specifically on whether the same glucose level indicates the same potential for disease in the various groups.

Cholesterol is the last laboratory measurement considered here in any detail. Data indicate that age and sex are major factors in cholesterol levels and that race differences may also exist. Cholesterol levels rise with age in both men and women, but their patterns of increase differ. In men, peak levels occur in the 45–54 age group, and then decline slightly thereafter (Abraham, Johnson, Carroll, 1978). In women the levels continue to rise steadily until age 65 and over, when the rise decelerates slightly. Figure 3.3 depicts cholesterol levels by age and sex. It indicates that in the 20s, 30s, and 40s men have higher cholesterol levels than women, but somewhere between the ages of 45 and 54, women's levels begin to exceed those of men.

United States Blacks and Whites have similar cholesterol levels (Abraham, Johnson, Carroll, 1978), but other groups differ substantially. Pima Indians from the southwestern United States and the Japanese have lower cholesterol levels than do United States Blacks or Whites (Hammon, Bennett, Hiller, 1975; Abraham, Johnson, Carroll, 1978). The mean value for Pima Indians is 189 mg/100 mL and for the Japanese is 152, while that for United States Blacks and Whites is 223. This mean value is not clinically

FIGURE 3.3 Mean serum cholesterol levels of adults 18–74 by sex: United States, 1971–1974

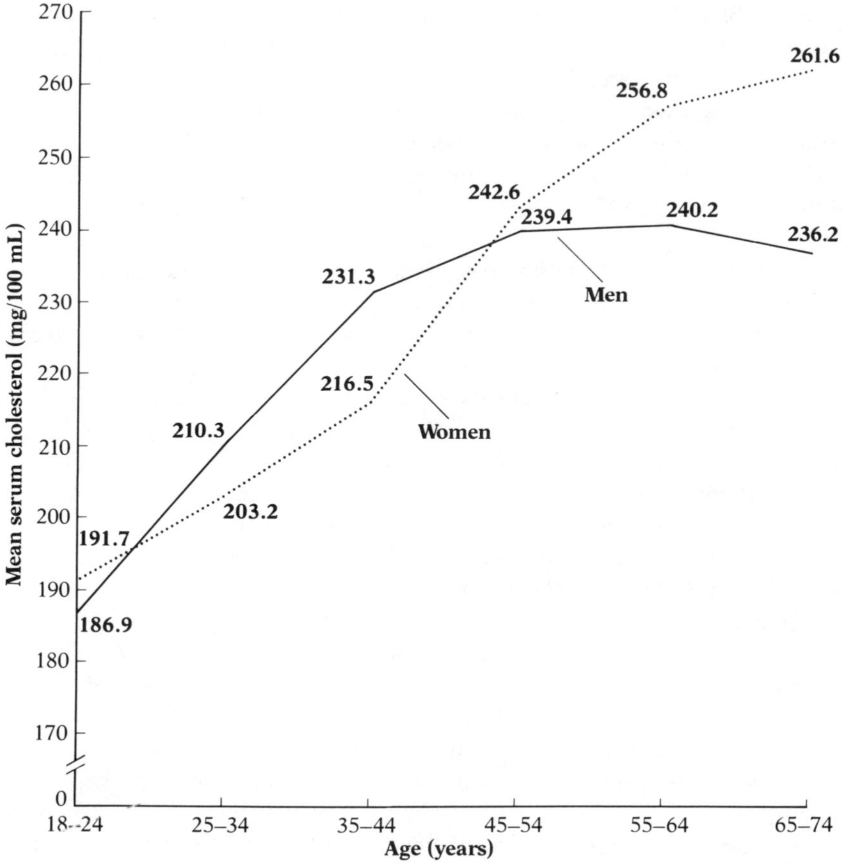

Source: Abraham et al, 1978, p 5.

meaningful because of large age and sex differences, but it does illustrate the magnitude of the differences between groups.

Because of differing cultures and food preferences it is not possible to determine if these differences are genetic. The fact that the Japanese and the Pima Indians belong to the same major racial group lends some credence to a racial difference in cholesterol levels.

Pulse and blood pressure are two of the more common measurements in the health field. Both may vary by age, and blood pressure may also vary by race. In healthy males 56 years of age, the pulse rate did not differ substantially from that taken 38 years earlier at age 18 (Damon, Roen, 1973): it decreased from 81.3 at age 18 to 76.3 at age 56. This difference is probably not clinically relevant, especially since the standard deviation is 12 beats for both measurement periods. A cross-sectional study of subjects from 20 to 90 years of age also produced no evidence of a decline in pulse rate with age (Luisada, Bhat, Knighten, 1980). The variability in normal pulse rates at all ages and the small size of the decline, if any, do not provide much useful information for clinicians.

It is well known that Blacks have higher blood pressure readings than Whites. While this statement is supported by data from the National Health Survey and other large population surveys done in the United States (Roberts, Maurer, 1977), there is reason to question the magnitude of the difference.

Blacks have an average blood pressure reading 5 mm Hg higher than Whites (Roberts, 1975), less at younger ages but more at older ages. Black women over 45 years have a reading as much as 16 mm Hg higher than White women in the same age group. At face value, these differences are statistically and perhaps clinically significant. But several other variables are involved that exaggerate the difference: education and income level, amount of body fat, and stress affect blood pressure (Gordon, 1964a; Florey, Acheson, 1969; Roberts, Maurer, 1977). It has not been satisfactorily determined how much of the Black-White blood pressure difference is due specifically to these factors. For example, stress occurs for many reasons, but the stress incurred during a physical examination was measured in one of the National Health Surveys by looking at the fall in blood pressure that occurred between the first and third measurement (Gordon, 1964a). Blood pressure decreased in both Blacks and Whites, but that of Blacks decreased 1 point more than in Whites. This explains 1 of the 5 points that differentiate the mean blood pressures of the two races. If all four of the variables—education, income, body fat, and stress—were considered, perhaps the blood pressure difference would disappear. Until such studies are done, a racial difference may be suspected, but its magnitude, even its existence, cannot be substantiated.

Other types of data support the notion that other variables are more important than race in determining blood pressure level. The increase in blood pressure that occurs when a group moves from a rural or very traditional environment to more urban and less traditional surroundings has been well documented (Finau, Prior, Evans, 1982). And a secular trend toward a decrease in blood pressure has also been noted in the United States (Roberts, 1975).

Disease Susceptibility

One of the definitions of aging mentioned earlier refers to an increasing vulnerability to disease and other bodily insults. This viewpoint supports separation of the causes of disease and disability in older individuals into three categories. The first is diseases that occur in varying degrees in all aging individuals. Atherosclerosis is a good example; it is universal and progressive, and it inevitably proves fatal if a more specific disease or injury does not intervene first. Osteoarthritis and osteoporosis are other examples. The second category includes diseases that are not universal but do increase in incidence with age. Diabetes, neoplasms, and heart disease are examples. The third category is diseases and bodily insults that affect people of all ages but have more serious consequences in older persons. Accidents and influenza fit into this category.

Of the nine leading causes of death in the United States, six are in the second category (diseases that increase in incidence with age): heart disease (the leading cause), cancer (second), cerebrovascular disease (third), chronic pulmonary disease (fifth), diabetes (seventh), and chronic liver disease (eighth) (National Center for Health Statistics, 1982). Two of the leading causes are in the third category (diseases with more serious consequences in the aged): accidents (fourth), and pneumonia and influenza (sixth). Only one is in the first category (diseases that occur in all aging individuals): atherosclerosis (ninth), accounting for only 1.5% of all deaths. The three leading causes (heart disease, cancer, and cerebrovascular disease) account for 68.3% of all deaths.

Since chapter 4 is devoted to disease, only two conditions connected with aging will be considered here: osteoarthritis and hypertensive heart disease. *Osteoarthritis* in the hands and feet affects 85% of people 75–79 years of age, as opposed to only 4% of those 18–24 (Roberts, Burch, 1966). Before age 45, men have a higher incidence of osteoarthritis than women, but after age 55, women have a higher incidence than men. There is a definite sex difference in the disease: women have more general involvement of all joints than men (Jurmain, 1980).

Data on racial differences are limited. Blacks and Whites appear to be similarly affected by osteoarthritis, but data on other groups are poor. The National Health Survey results suggest that Orientals and American Indians have less osteoarthritis in the hands and feet than Blacks and Whites do (Roberts, Burch, 1966).

Heavy joint use and severe stress to specific joints are probably more important than race, and possibly more important than sex, in explaining the presence and severity of osteoarthritis. The disease is more prevalent in heavy individuals than in normal-weight controls (Engel, 1968; Dequeker, Goris, Uytterhoeven, 1983). It occurs more often in men who work at heavy manual labor than in men with less physically demanding jobs (Engel, Burch, 1966). A study of skeletal remains of four racial groups—modern Blacks and Whites and protohistoric Pima Indians and Eskimos—indicated that osteoarthritic involvement of particular joints correlated well with the types of physical activities engaged in by the groups (Jurmain, 1977).

The higher risk of *hypertensive heart disease* in Blacks, compared with Whites, is as well known as their higher blood pressure levels. But the same question arises here as for blood pressure: are the differences biologic or socioeconomic? Among Blacks, hypertensive heart disease becomes less common as educational and income levels rise (Gordon, Devine, 1973). The rate is also substantially lower than expected in employed Black men compared with those who are retired or of unknown employment level. Black men in professional, technical, and managerial positions, those who are independent farmers or farm managers, and those who are draftsmen and foremen have much lower rates of hypertensive heart disease than Black men in lower-status jobs. Similar but less pronounced differences apply to Black women. Male Black workers in government transportation, communication, and public utility jobs have much reduced rates of hypertensive heart disease compared with their counterparts in other industries. Such data considerably weaken the biologic explanation for different rates of hypertensive heart disease between Blacks and Whites.

Longevity Prediction and Promotion

Numerous attempts have been made to devise a test battery to determine physiologic age as opposed to calendar age. Such tests would be useful for predicting longevity, for evaluating exercise and dietary prescriptions that purport to extend life, and for assessing environmental hazards that might reduce life expectancy (Comfort, 1969; Palmore, 1969; Borkan, Norris, 1980). The tests generally consist of several discrete biological and intellectual measurements that are standardized by age. For instance, the mean blood pressure, reaction time, and skin elasticity values for specific age groups are calculated. An individual's measurements are then compared with the age standards. Several of these tests correlate very well with calendar age (Comfort, 1969; Borkan, Norris, 1980). But an examining physician's simple visual inspection and estimation of age was found to compare closely to the more complicated age prediction tests (Borkan, Bachman, Norris, 1982). The physicians were not instructed on how to estimate age, which suggests that, at least within the racial group of the examiner, age estimation is fairly accurate. However, there was a systematic error in estimation of the very oldest subjects; those over 75 years of age were consistently estimated to be younger than they were. The investigators thought that this occurred either because the subjects over 75 were healthy enough to come to the examining center, while less healthy oldsters had dropped out of the study, or because the physicians lacked experience with healthy old people.

Both the visual estimation of age and the more complicated tests for age prediction were closely correlated with mortality. Those who looked or measured older than their actual age died sooner than those who looked their age or younger (Borkan, Bachman, Norris, 1982). Both age measures also indicated that individuals who were estimated or predicted to be younger than their actual age were physically more active than those who looked their age or looked older. Whether physical activity keeps people young or whether very healthy people are the ones who keep active in old age is still not completely clear.

Although the roles of physical exercise, diet, and other practices in the maintenance of health into old age are not well understood, many reports suggest such influences. People who maintain fairly vigorous physical activity are healthier and live longer than their age peers who do not exercise very much (DeVries, 1970; Belloc, Breslow, 1972; Adams, DeVries, 1973; Belloc, 1973). The association between exercise status, health, and mortality is well established, but whether it is causal is still debated. Longitudinal studies are necessary to answer the question, and they are hard to do. The few studies that have been done indicate that exercise improves health, but they give little evidence that it prolongs life (DeVries, 1970; Adams, DeVries, 1973; Kasch et al, 1973; Sidney, Shephard, 1977). Improved health, of course, is a worthwhile goal in itself, even if life is not prolonged. Quality of life is more important than mere quantity.

Regular physical exercise has been shown to increase the efficiency of the respiratory and cardiovascular systems (Kasch et al, 1973; Adams, DeVries, 1973; Sidney, Shephard, 1977). It promotes fat metabolism (Wood et al, 1976) and halts or slows the total body calcium and potassium losses that are generally seen in aging (Aloia, 1981). Besides these and other physical improvements, it offers psychologic and social benefits.

The role of diet in increasing health and promoting longevity is even less well documented than the role of exercise. People who eat breakfast daily and people who refrain from habitual snacks are healthier than those who skip breakfast and those who snack between meals (Belloc, Breslow, 1972). Moderate alcohol drinkers and abstainers are healthier than people who drink heavily (more than five drinks at a time). People who are more than 20% overweight are less healthy than their thinner cohorts. These diet-related attributes pose the same interpretation problem as the exercise attributes: which comes first, the good health or the good diet?

Dietary controversies and fads related to cholesterol, fiber, high or low protein, refined or complex carbohydrates, and the like are well known. Until science settles these matters, people can and should eat as they wish, within moderation, of course. The fact that so many different cuisines with different basic foodstuffs have evolved that all support life adequately gives me hope that the current morbid concern over dangerous food items and patterns will ultimately abate.

Two other practices that affect health are intellectual activity and sleeping patterns. Individuals who sleep 7 or 8 hours per night are healthier and live longer than those who sleep less or more than that (Belloc, Breslow, 1972; Belloc, 1973). Intellectual activity, like physical activity, is also strongly correlated with health. People who are more physically active are also more intellectually active, and they score higher than the physically inactive on measures of intelligence. Which activity influences which is not known. Perhaps they are both due to another variable, heredity, for there is evidence that long-lived individuals come from long-living families. Again, is this biologic, or are the health practices of the family taught to the next generation?

Summary

This section has focused primarily on the normal aging process. The amount of data available on healthy old people is sparse compared to the abundance on sick old people. The focus on illness rather than health is also apparent among health care providers, who come in contact mostly with the institutionalized, hospitalized, or home-bound sick old people. "Why are we so slow in establishing . . . health care systems for the elderly? Moreover, why do we know so little about the health of the most rapidly growing segment of our population?" (Gelein, 1982, p 180). Health care practitioners need to put more effort into helping the 95% of healthy elderly stay healthy.

The best way to ensure that people live—really live, not just exist—is to encourage them to be physically and mentally active, to get 7 to 8 hours of sleep a night, to eat breakfast regularly and not to overeat or overindulge in alcohol. If people want to look younger, they need to be born to Black, Oriental, or Native American parents rather than White parents, because Whites turn gray and wrinkle the soonest.

As people age, they get a little shorter and somewhat broader around the middle. They lose some muscle and bone mass but do not weigh much less until their 60s (males) or 70s (females) because of a gain in fat. Their eyes lose ability to accommodate, so glasses are needed by the middle to late 40s. Most keep their hearing, ability to savor good food, and sense of touch and pain. From the wear and tear of life, they ususally incur some joint stiffness (osteoarthritis) and some atherosclerosis but are not incapacitated. They may become more susceptible to the chronic diseases and some infectious diseases but can often avoid these by keeping up healthy habits. With some luck and good planning, they may die of old age before anything worse comes along to end their lives.

REFERENCES

Growth and Development

Adams MS, Niswander JD: Birth weight of North American Indians: A correlation and amplification. *Hum Biol* 1973; 45(3):351–357.

Ashworth A: Growth rates in children recovering from protein calorie malnutrition. *Br J Nut* 1969; 23:835–844.

Barr GD, Allen CM, Shinefield HR: Height and weight of 7,500 children of three skin colors. *Am J Dis Child* 1972; 124:866–872.

Batson R, Young WC, Shepard FM: Observations on skin resistance to electricity and sweat chloride content. *J Pediatr* 1962; 60(5):716–720.

Bayley N: Comparisons of mental and motor test scores for ages 1–15 months by sex, birth order, race, geographical location, and education of parents. *Child Dev* 1965; 36:379–411.

Beall C et al: The effects of high altitude on adolescent-growth in southern Peruvian Amerindians. *Hum Biol* 1977; 49(2):109–124.

Blanco RA et al: Height, weight and lines of arrested growth in young Guatemalan children. *Am J Phys Anthropol* 1974; 40(1):39–48.

Brody JA, Overfield T, McAlister R: Draining ears and deafness among Alaskan Eskimos. *Arch Otolaryngol* 1965; 81:29–33.

Brothwell D: *Dental Anthropology*. Pergamon Press, 1963.

Bucher KA et al: Racial difference in incidence of ABO hemolytic disease. *Am Public Health* 1976; 66(9):854–858.

Bulmer M: *The Biology of Twinning in Man*. Oxford University Press, 1970.

Campbell, JJ, Oski, FA: Sickle cell anemia in an American White boy of Greek ancestry. *Am J Dis Child* 1977; 131:186–188.

Caskey, DA et al: Effects of age on lactose malabsorption in Oklahoma Native Americans as determined by breath H_2 analysis. *Digestive Dis* 1977; 22(2):113–116.

Cavalli-Sforza LL, Bodmer, WF: *The Genetics of Human Populations*. Freeman, 1971.

Centers for Disease Control: *Nutrition Surveillance*. HHS Publication No. (CDC) 81–8295. US Department of Health and Human Services, Public Health Service, January–June 1980.

Chagula, WK: The age at eruption of third permanent molars in male East Africans. *Am J Phys Anthropol* 1960; 18(1):77–82.

Ching GHS, Ching CS, Nemechek RW: Genetic and epidemiological studies of clubfoot in Hawaii: Ascertainment and incidence. *Am J Hum Genet* 1969; 21:566–580.

Christianson RE et al: Incidence of congenital anomalies among White and Black live births with long-term follow-up. *Am J Public Health* 1981; 71(12):1333–1339.

Chung CS, Myrianthopoulos NC: Racial and prenatal factors in major congenital malformations. *Am J Hum Genet* 1968; 20:44–60.

Diamond LK: Racial inequality and neonatal jaundice. *Am J Public Health* 1976; 66(9):835–836.

Dover GJ, Boyer SH, Pembrey, ME: F-cell production in sickle cell anemia: Regulation by genes linked to β-hemoglobin locus. *Science* 1981; 211(4489):1441–1444.

Duncan B, Smith AN, Briese FW: A comparison of growth: Spanish surnamed with non-Spanish-surnamed children. *Am J Public Health* 1979; 69(9):903–907.

Ellis WS: Mortality and birth weight in Philadelphia blacks: An example of stabilizing selection. *Am J Phys Anthropol* 1972; 38(1):145–150.

Emanuel I et al: The incidence of congenital malformations in a Chinese population: The Taipei Collaborative Study. *Teratology* 1972; 5(2): 159–169.

Erickson JD: Racial variations in the incidence of congenital malformations. *Ann Hum Genet* 1976; 39:315–320.

Ershoff BH: Comparative microhematocrit values of Spanish Americans and Whites attending Los Angeles County Youth Clinics. *Public Health Rep* 1975; 90(1):58–61.

Eveleth P: The effects of climate on growth. *Ann NY Acad Sci* 1966; 134(2):750–759.

Eveleth P, Tanner J: *Worldwide Variation in Human Growth*. Cambridge University Press, 1976.

Falkner F, Tanner J: *Human Growth I—Principles and Prenatal Growth.* Plenum Press, 1978.

Farrell PM, Wood RE: Epidemiology of hyaline membrane disease in the United States: Analysis of national mortality statistics. *Pediatrics* 1976; 58(2):167–176.

Fisher O et al: Jaundice and breast-feeding among Alaskan Eskimo newborns. *Am J Dis Child* 1978; 132:859–861.

Foster TA et al: Anthropometric and maturation measurements of children, ages 5 to 14 years, in a biracial community—the Bogalusa Heart Study. *Am J Clin Nutr* 1977; 30:582–591.

Frankenburg WK, Dodds JB: The Denver Developmental Screening Test. *J Pediatr* 1967; 71(2):181–191.

Freitas M, Salazano FM: Eruption of permanent teeth in Brazilian Whites and Blacks. *Am J Phys Anthropol* 1975; 42(1):145–150.

Frisancho RA: Developmental responses to high altitude hypoxia. *Am J Phys Anthropol* 1970; 32(3):401–408.

Frisancho RA et al: Adaptive significance of small body size under poor socioeconomic conditions in southern Peru. *Am J Phys Anthropol* 1973; 39(2):255–262.

Fuller GN, Friedrich D: Visual motor test performance: Race and achievement variables. *J Clin Psychol* 1979; 35(6):621–630.

Fulwood R, Abraham S, Johnson C: Serum cholesterol levels of persons 4–74 years of age by socioeconomic characteristics. *Vital Health Stat* 1980; 11(217):1–30.

Garn SM, Clark DC, Guire, KE: Level of fatness and size attainment. *Am J Phys Anthropol* 1974; 40(3):447–449.

Garn SM, Smith JJ, Clark DC: The magnitude and the implications of apparent race differences in hemoglobin values. *Am J Clin Nutr* 1975; 28:563–568.

Garn SM et al: Negro-Caucasoid differences in permanent tooth emergence at a constant income level. *Arch Oral Biol* 1973; 18:609–615.

Glueck CJ et al: Black-White similarities in cord blood lipids and lipoproteins. *Metabolism* 1977; 26(4):347–350.

Greene LS: Physical growth and development, neurological maturation and behavioral functioning in two Ecuadoran Andean communities in which goiter is epidemic. *Am J Phys Anthropol* 1974; 41(1):139–152.

Griffith TE: Epidemiology of otitis media—an interracial study. *Laryngoscope* 1979; 89:22–29.

Haas JD et al: Altitude, ethnic and sex differences in birth weight and length in Bolivia. *Hum Biol* 1980; 52(3):459–477.

Hamill PV, Johnston FE, Lemeshow S: Body weight, stature, and sitting height: White and Negro youths 12–17 years. *Vital Health Stat* 1973; 11(126):1–27.

Hamill PV et al: Physical growth: National Center for Health Statistics percentiles. *Am J Clin Nutr* 1979; 32:607–629.

Hardy JB, Mellits ED: Relationship of low birth weight, gestational age, placental weight and mortality. Pages 105–118, in: Reed DM, Stanley FJ (editors). Urban and Schwarzenberg, 1977.

Heald F et al: Hematocrit values of youths 12–17 years. *Vital Health Stat* 1974; 11(146):1–16.

Heller CA, Scott EM, Hammes LA: Height, weight, and growth of Alaskan Eskimos. *Am J Dis Child* 1967; 113:338–344.

Horiguchi T, Bauer C: Ethnic differences in neonatal jaundice: Comparison of Japanese and Caucasian newborn infants. *Am J Obstet Gynecol* 1975; 121(1):71–74.

Jacobs AH, Walton RG: Incidence of birthmarks in the neonate. *Pediatrics* 1976; 58:218–222.

Jarvis A, Gorlin RJ: Minor orofacial abnormalities in Eskimo population. *Oral Surg* 1972; 33:417–427.

Johnson CL, Abraham S: Hemoglobin and selected iron related findings of persons 1–74 years of age: United States, 1971–74. *Advancedata* 1979; 46:1–11.

Johnson RC, Cole RE, Abua FM: Genetic interpretation of racial/ethnic differences in lactose absorption and tolerance: A review. *Hum Biol* 1981; 53(1):1–13.

Johnson RL: Chronic otitis media in school age Navajo Indians. *Laryngoscope* 1967; 77:1990–1995.

Johnston FE, Hamill PV, Lemeshow S: Skinfold thickness of children 6–11 years, United States. *Vital Health Stat* 1972; 11(120):1–17.

Johnston FE et al: Physical growth and development of urban Native Americans: A study in urbanization and its implication for nutritional status. *Am J Clin Nutr* 1978; 31:1017–1027.

Kimura K: Skeletal maturation of children in Okinawa. *Ann Hum Biol* 1976; 3(2):149–155.

Kirkman HN: Further evidence for a racial difference in frequency of ABO hemolytic disease. *J Pediatr* 1977; 90(5):717–721.

Lauritsen JG, Grunnet N, Jensen OM: Materno-fetal ABO incompatibility as a cause of spontaneous abortion. *Clin Genet* 1975; 7:308–316.

Lebenthal E, Antonowicz I, Schwochman H: Correlation of lactose activity, lactose intolerance, and milk consumption in different age groups. *Am J Clin Nutr* 1975; 28:595–600.

Lechtig A et al: Effect of food supplementation during pregnancy on birthweight. *Pediatrics* 1975; 56(4):508–519.

Levy PS et al: Total serum cholesterol values of youths 12–17 years. *Vital Health Stat* 1976; 11(156):1–11.

Lichty JA et al: Studies of babies born at high altitude. *Am J Dis Child* 1957; 93:666–678.

Lynds BG, Seyler SK, Morgan BM: The relationship between elevated blood pressure and obesity in Black children. *Am J Public Health* 1980; 70(2):171–173.

MacMahan B: Age at menarche. *Vital Health Stat* 1973; 11(133):1–7.

Malina R: Skeletal maturation studied longitudinally over one year in American Whites and Negroes, six through thirteen years of age. *Hum Biol* 1970; 42(3):377–390.

Malina RM, Hamill PV, Lemeshow S: Body dimensions and proportions, White and Negro children 6–11 years. *Vital Health Stat* 1974; 11(143):1–47.

Manning P, Avery ME: Purulent otitis media differences between populations in different environments. *Pediatrics* 1974; 53(2):135–136.

Martin J, Crump P: Leukoedema of the buccal mucosa in Negro children and youth. *Oral Surg* 1972; 34(1):49–58.

Martinek TJ, Cheffers JT, Zaichkowsky LD: Physical activity, motor development and self-concept: Race and age differences. *Percept Mot Skills* 1978; 46:147–154.

Martorell R et al: Malnutrition, body size, and skeletal maturation: Interrelationships and implications for catch-up growth. *Hum Biol* 1979; 51(3):371–389.

Matsunaga E: The dimorphism in human normal cerumen. *Ann Hum Genet* 1962; 25:273–286.

McGrew CJ: Sickle cell trait in the White population. *JAMA* 1973; 224(13):1762–1763.

Meredith HV: *Human Body Growth in the First Ten Years of Life*. State Printing Co., 1978.

Morton NE: Genetic aspects of prematurity. Pages 213–231 in: *The Epidemiology of Prematurity*. Reed DM, Stanley FJ (editors). Urban & Schwarzenberg, 1977.

Nankin HR et al: Correlation between sexual maturation and serum gonadotropins: Comparison of Black and White youngsters. *Am J Med Sci* 1974; 268(3):139–147.

National Center for Health Statistics: Advance report of final natality statistics, 1979. *Monthly Vital Stat Rep* 1981; 30(6):1–27.

Newcomer AD et al: Family studies of lactose deficiency in the American Indian. *Gastroenterology* 1977; 73(5):985–988.

Niswander JD, Adams MS: Oral clefts in the American Indian. *Public Health Rep* 1967; 82(9):807–812.

Olowe SA, Akinkugbe A: Amniotic fluid lecithin/sphingomyelin ratio: Comparison between an African and a North American community. *Pediatrics* 1978; 62(1):38–41.

Palomino H, Mueller WH, Schull WJ: Altitude, heredity, and body proportion in Northern Chile. *Am J Phys Anthropol* 1979; 51(1):39–50.

Peevy KJ, Wiseman HJ: ABO hemolytic disease of the newborn: Evaluation of management and identification of racial and antigenic factors. *Pediatrics* 1978; 61(3):475–478.

Postl BD, Nelson N, Carson J: Hyperbilirubinemia in Inuit neonates. *Can Med Assoc J* 1982; 126:811–813.

Pratt MW, Janus ZL, Sayal NC: National variations in prematurity (1973 and 1974). Pages 53–79 in: *The Epidemiology of Prematurity*. Reed DM, Stanley FJ (editors). Urban & Schwarzenberg, 1977.

Querec LJ: Apgar score in the United States, 1978. *Monthly Vital Stat Rep* 1981; 30(1):1–15.

Rauh J, Schumsky DA: Relative accuracy of visual assessment of juvenile obesity. *J Am Diet Assoc* 1969; 55:459–464.

Reed D, Struve S, Maynard JE: Otitis media and hearing deficiency among Eskimo children: A cohort study. *Am J Public Health* 1967; 57(9):1657–1662.

Reed WL: Racial differences in blood pressure levels of adolescents. *Am J Public Health* 1981; 71(10):1165–1167.

Ritchie RF: Specific proteins. Pages 228–258 in: *Clinical Diagnosis and Management by Laboratory Methods*. Henry JB (editor). Saunders, 1979.

Roche AF, Roberts J, Hamill PV: Skeletal maturity of youths 12–17 years: Racial, geographic, and socioeconomic differentials. *Vital Health Stat* 1978; 11(167):1–98.

Rode A, Shephard, RJ: Cardiac output, blood volume and total hemoglobin of the Canadian Eskimo. *J Appl Physiol* 1973; 34(1):91–96.

Roode H et al: Serum transferrin values in White and Black toddlers. *South Afr Med J* (March) 1975; 49(9):319–321.

Rowland M, Roberts J: Blood pressure levels and hypertension in persons ages 6–74 years: United States, 1976–1980. *Vital Health Stat* 1982; 84:1–12.

Rush D, Stein Z, Susser M: A randomized controlled trial of prenatal nutritional supplementation in New York City. *Pediatrics* 1980; 65(4):683–697.

Saland J, McNamara H, Cohen MJ: Navajo jaundice: A variant of neonatal hyperbilirubinemia associated with breast feeding. *J Pediatr* 1974; 85(2):271–275.

Savage PJ et al: Serum cholesterol levels in American (Pima) Indian children and adolescents. *Pediatrics* 1976; 58(2):274–282.

Schachter J, Lachin JM, Wimberly FD: Newborn heart rate and blood pressure: Relation to race and to socioeconomic class. *Psychosom Med* 1976; 38(6):390–397.

Schutte JE: Growth differences between lower and middle income black male adolescents. *Hum Biol* 1980; 52(2):193–204.

Scott RB et al: Growth and development of Negro infants: V. Neuromuscular patterns of behavior during the first year of life. *Pediatrics* 1955; 16(24):24–29.

Timmermans FJ, Gerson S: Chronic granulomatous otitis media in bottle fed Inuit children. *Can Med Assoc J* 1980; 122:545–547.

Villa-Dresser CM, Carroll MD, Abraham S: Selected findings: Food consumption profiles of White and Black persons, 1–7 years of age in the United States, 1971–1974. *Advancedata* 1978; 21:1–11.

Walker ARP et al: Blood pressures of Indian and colored school children aged 10–12 years. *S Afr Med J* (August) 1978; 54(6):315–318.

Weiner JS, Lourie JA: *Human Biology: A Guide to Field Methods*. IBP Handbook No. 9. Blackwell Scientific Publications, 1969.

Whitner W, Thompson M: The influence of bathing on the newborn infant's body temperature. *Nurs Res* 1970; 19(1):30–36.

Williams JR, Scott RB: Growth and development of Negro infants: IV. Motor development and its relationship to child rearing practices in two groups of Negro infants. *Child Dev* 1953; 24(2):103–121.

World Health Organization: *A Growth Chart for International Use in Maternal and Child Health Care*. WHO, 1978.

Yanocheck-Owen A, White M: Nutrition surveillance in Arizona: Selected anthropometric and laboratory observations among Mexican-American children. *Am J Public Health* 1977; 67(2):151–154.

Aging

Abraham S, Johnson CL, Carroll MD: Total serum cholesterol levels of adults 18–74 years. *Vital Health Stat* 1978; 11(205):1–31.

Abraham S, Johnson CL, Najjar MF: Height and weight of adults 18–74 years of age in the United States. *Advancedata* 1976; (3):1–7.

Abraham S, Johnson CL, Najjar MF: Weight by height and age of adults 18–74 years: United States, 1971–74. *Advancedata* 1977; (14):1–11.

Adams GM, DeVries HA: Physiological effects of an exercise training regimen upon women aged 52–79. *J Gerontol* 1973; 28(1):50–55.

Admundsen D, Diers C: The age of menopause in medieval Europe. *Hum Biol* 1973; 45(4):505–512.

Aloia JF: Exercise and skeletal health. *J Am Geriatr* 1981; 29(3):104–107.

Arnetz BB, Anders K, Tores T: The influence of aging on hemoglobin A_{1c} (HbA_{1c}). *J Gerontol* 1982; 37(6):648–650.

Belloc NB: Relationship of health practices and mortality. *Prev Med* 1973; 2:67–81.

Belloc NB, Breslow L: Relationship of physical health status and health practices. *Prev Med* 1972; 1:409–421.

Borkan GA, Bachman SS, Norris AH: Comparison of visually estimated age with physiologically predicted age as indicators of rates of aging. *Soc Sci Med* 1982; 16(2):204.

Borkan GA, Norris AH: Assessment of biological age using a profile of physical parameters. *J Gerontol* 1980a; 35(2):177–184.

Borkan GA, Norris AH: Biological age in adulthood: Comparison of active and inactive U.S. males. *Hum Biol* 1980b; 52(4):787–802.

Borkan GA, Norris AH: Fat redistribution and the changing body dimensions of the adult male. *Hum Biol* 1977; 49(3):495–514.

Brand PC, Lehert PH: A new way of looking at environment variables that may affect the age at menopause. *Maturitas* 1978; 1:121–132.

Brues AM: *People and Races*. Macmillan, 1977.

Burch PRJ, Murray JJ, Jackson D: The age prevalence of arcus senilis, greying of hair and baldness, etiological considerations. *J Gerontol* 1971; 26(3):364–372.

Burnham CE: Edentulous persons. *Vital Health Stat* 1974; 10(89):1–31.

CECOS Federation, Schwartz D, Mayaux MJ: Female fecundity as a function of age. *N Engl J Med* 1982; 306(7):404–406.

Clement FJ: Longitudinal and cross sectional assessments of age changes in physical strength as related to sex, social class, and mental ability. *J Gerontol* 1974; 29(4):423–429.

Comfort A: Test-battery to measure aging-rate in man. *Lancet* (December 27) 1969; 2(7635):1411–1415.

Damon A, Roen JL: Aging in the Solomon Islands and the United States: Tests of Pearl's hypothesis. *Hum Biol* 1973; 45(4):683–693.

Damon A et al: Age and physique in healthy White veterans at Boston. *J Gerontol* 1972; 27(2):202–208.

Dequeker J, Goris P, Uytterhoeven R: Osteoporosis and osteoarthritis (osteoarthrosis): Anthropometric distinctions. *JAMA* 1983; 249(11):1448–1451.

Devine B: Mean blood hematocrit of adults. *Vital Health Stat* 1967; 11(24):1–36.

DeVries HA: Physiological effects of an exercise training regimen upon men aged 52–58. *J Gerontol* 1970; 25(4):325–336.

Engel A: Osteoarthritis and body measurements. *Vital Health Stat* 1968; 11(29):1–37.

Engel A, Burch TA: Osteoarthritis in adults by selected demographic characteristics. *Vital Health Stat* 1966; 11(20):1–27.

Ericksen MF: Aging changes in the medullary cavity of the proximal femur in American Blacks and Whites. *Am J Phys Anthropol* 1979; 51(4):563–570.

Ericksen MF: Aging in the lumbar spine: II. L1 and L2. *Am J Phys Anthropol* 1978a; 48(2):241–246.

Ericksen MF: Aging in the lumbar spine: III. L5. *Am J Phys Anthropol* 1978b; 48(2):247–250.

Ericksen MF: Cortical bone loss with age in three Native American populations. *Am J Phys Anthropol* 1976; 45(3):443–452.

Finau SA, Prior IA, Evans JG: Aging in the South Pacific—physical changes with urbanization. *Soc Sci Med* 1982; 16:1539–1549.

Florey CduV, Acheson RM: Blood pressure as it relates to physique, blood glucose and serum cholesterol. *Vital Health Stat* 1969; 11(34):1–29.

Forbes GB: The adult decline in lean body mass. *Hum Biol* 1976; 48(1):161–173.

Frere G: Mean age of menopause and menarche in South Africa. *S Afr J Med Sci* 1971; 36:21.

Friedlaender JS et al: Longitudinal physique changes among healthy White veterans at Boston. *Hum Biol* 1977; 49(4):541–558.

Fries JF: Aging, natural death, and the compression of morbidity. *N Engl J Med* 1980; 303(3):130–135.

Garn S, Pao EM, Rihl ME: Compact bone in Chinese and Japanese. *Science* 1964; 143(3613):1439–1440.

Garry PJ, Goodwin JS, Hunt WC: Iron status and anemia in the elderly: New findings and a review of previous studies. *J Am Geriatr Soc* 1983; 31(7):389–399.

Garst CC: Blood glucose level in adults. *Vital Health Stat* 1966; 11(18):1–25.

Gelein JL: Aged women and health. *Nurs Clin North Am* 1982; 17(1):179–185.

Gentile A: Persons with impaired hearing. *Vital Health Stat* 1975; 10(101):1–43.

Gilchrest BA: Age-associated changes in the skin. *J Am Geriatr Soc* 1982; 30(2):139–143.

Gordon T: Blood pressure of adults by race and area. *Vital Health Stat* 1964a; 11(5)1–20.

Gordon T: Glucose tolerance of adults. *Vital Health Stat* 1964b; 11(2):1–25.

Gordon T, Devine B: Hypertension and hypertensive heart disease in adults. *Vital Health Stat* 1973; 11(13):1–62.

Grzegorczyk PB, Jones SW, Mistretta CM: Age-related differences in salt taste acuity. *J Gerontol* 1979; 34(6):834–840.

Ham AW: *Histology*, 7th ed. Lippincott, 1974.

Hammon RF, Bennett PH, Hiller M: The effect of menopause on serum cholesterol in American (Pima) Indian women. *Am J Epidemiol* 1975; 102(2):164–169.

Hatton J: Aging and the glare problem. *J Gerontol Nurs* 1977; 3(5):38–44.

Hayflick L: The biology of human aging. *Plast Reconstr Surg* 1981; 67(4):536–550.

Hollingsworth JW, Hashizume A, Jablon S: Correlations between tests of aging in Hiroshima subjects—an attempt to define "physiologic age." *Yale J Biol Med* 1965; 38:11–26.

Israel H: The dichotomous pattern of craniofacial expansion during aging. *Am J Phys Anthropol* 1977; 47(1):47–52.

Jernigan JA et al: Reference values for blood findings in relatively fit elderly persons. *J Am Geriatr Soc* 1980; 28(7):308–314.

Jurmain RD: The pattern of involvement of appendicular degenerative joint disease. *Am J Phys Anthropol* 1980; 53(1):143–150.

Jurmain RD: Stress and the etiology of osteoarthritis. *Am J Phys Anthropol* 1977; 46(2):353–366.

Karawawa A, Kawashima K, Kasahara H: Mental aging and its medico-psycho-social background in the very old Japanese. *J Gerontol* 1979; 34(5):680–686.

Kasch FW et al: Cardiovascular changes in middle-aged men during two years of training. *J Appl Physiol* 1973; 34(1):53–57.

Keogh EV, Walsh RJ: Rate of greying of human hair. *Nature* 1965; 207:877–878.

Kligman AM: Perspectives and problems in cutaneous gerontology. *J Invest Dermatol* 1979; 73(1):39–46.

Kohn RR: Cause of death in very old people. *JAMA* 1982; 247(20):2793–2797.

Leaf A: Long-lived populations: Extreme old age. *J Am Geriatr* 1982; 30(8):485–487.

Luisada AA, Bhat PK, Knighten V: Changes in cardiac output caused by aging: An impedance cardiographic study. *Angiology* 1980; 31(2):75–81.

MacMahon B, Worcester J: Age at menopause. *Vital Health Stat* 1966; 11(19):1–20.

Mazess RB, Mather WE: Bone mineral content in Canadian Eskimos. *Hum Biol* 1975; 47(1):45–63.

Metropolitan Life Insurance Company: Expectations of life among non-Whites. *Stat Bull Metropol Life Ins Co* 1977; 55(1):1–7.

Metropolitan Life Insurance Company: Expectations of life in the U.S. at new high. *Stat Bull Metropol Life Ins Co* 1980; 61(4):13–15.

Moore LM, Nielsen CR, Mistretta CM: Sucrose taste thresholds: Age related differences. *J Gerontol* 1982; 37(1)64–69.

Morgan DB et al: Amount of bone in the metacarpal and the phalanx according to age and sex. *Clin Radiol* 1967; 18:101–108.

National Center for Health Statistics: Advance report of final mortality statistics, 1979. *Monthly Vital Stat Rep* 1982; 31(6):1–57.

National Center for Health Statistics: Annual summary of births, deaths, marriages, and divorces: United States, 1980. *Monthly Vital Stat Rep* 1981; 29(13):1–9.

Oland LM, Mason RL, Alexeff AI: Bone density and dietary findings of 409 Tennessee subjects: I. Bone density considerations. *Am J Clin Nutr* 1972; 25:905–907.

Overfield T, Call EB: Earlobe type, race, and age: Effects on earlobe creasing. *J Am Geriatric Soc* 1983; 31(8):479–481.

Palmore E: Predicting longevity: A follow-up controlling for age. *Gerontologist* (Winter) 1969; 9(4):247–251.

Parizkova J, Eiselt E: A further study on changes in somatic characteristics and body composition of old men followed longitudinally for 8–10 years. *Hum Biol* 1971; 43(3):318–326.

Parizkova J, Eiselt E: Longitudinal changes in body build and skinfolds in a group of old men over a 16 year period. *Hum Biol* 1980; 52(4):803–809.

Pfeiffer E: Survival in old age: Physical, psychological and social correlates of longevity. *J Am Geriatr Soc* 1970; 18(4);273–285.

Pfeiffer S: Age changes in the external dimensions of adult bone. *Am J Phys Anthropol* 1980; 52(4):529–532.

Pochi PE, Strauss JS, and Downing DT: Age-related changes in sebaceous gland activity. *J Invest Dermatol* 1979; 73(1):108–111.

Potvin AR et al: Human neurologic function and the aging process. *J Am Geriatric Soc* 1980; 28(1):1–9.

Roberts J: Binocular visual acuity of adults. *Vital Health Stat* 1964; 11(3):1–27.

Roberts J: Blood pressure of persons 18–74 years. *Vital Health Stat* 1975; 11(150):1–23.

Roberts J, Burch TA: Osteoarthritis prevalence in adults by age, sex, race, and geographic area. *Vital Health Stat* 1966; 11(15):1–27.

Roberts J, Cohrssen J: History and examination findings related to visual acuity among adults. *Vital Health Stat* 1968; 11(28):1–31.

Roberts J, Maurer K: Blood pressure levels of persons 6–74 years. *Vital Health Stat* 1977; 11(203):1–103.

Rowland M: Basic data on hearing levels of adults 25–74 years. *Vital Health Stat* 1980; 11(25):1–49.

Schludermann E, Zubek JP: Effect of age on pain sensitivity. *Percept Mot Skills* 1962; 14:295–301.

Sherman DE, Robillard E: Sensitivity to pain in relationship to age. *J Am Geriatric Soc* 1964; 12:1037–1044.

Sidney KH, Shephard RJ: Activity patterns of elderly men and women. *J Gerontol* 1977; 32(1):25–32.

Szabo G: The skin as an adaptive organ. Pages 39–58 in: *Physiological Anthropology.* Damon A (editor). Oxford University Press, 1975.

Thornbury JM, Mistretta CM: Tactile sensitivity as a function of age. *J Gerontol* 1981; 36(1):34–39.

Treloar A: Menarche, menopause and intervening fecundability. *Hum Biol* 1974; 46(1):89–107.

Tripathi RC, Tripathi BJ: Lens morphology, aging, and cataract. *J Gerontol* 1983; 38(3):258–270.

Trotter M, Gleser G: The effect of aging on stature. *Am J Phys Anthropol* 1951; 9:311–324.

Trotter M, Peterson RR: Some variable factors in the adult skeleton. *Ann NY Acad Sci* 1966; 134(2):841–845.

Verrillo RT: Age related changes in the sensitivity to vibration. *J Gerontol* 1980; 35(2):185–193.

Weiffenbach JM, Baum BJ, Burghauser R: Taste thresholds: Quality specific variation with human aging. *J Gerontol* 1982; 37(3):372–377.

Winterer JC et al: Whole body protein turnover in aging man. *Exp Gerontol* 1976; 11:79–87.

Wood PD et al: The distribution of plasma lipoproteins in middle-aged male runners. *Metabolism* 1976; 25(11):1249–1255.

chapter 4

Biochemical variation and differential disease susceptibility

Readers may ask, why are biochemical variation and disease susceptibility discussed in the same chapter, and why consider disease in a book on normal biologic variation? The answer to the first question is that the more health professionals know about specific diseases, the more molecular their knowledge becomes. For instance, alcoholism is not caused exclusively by psychosocial factors; researchers now know that a biologic factor, namely, a specific difference in acetaldehyde metabolism, contributes to alcohol dependence.

This book on normal variation discusses disease partly because disease is almost a normal occurrence in humans and partly because many common diseases show racial variation. Many other diseases also differ by group, but only the more common diseases, those that affect large numbers of people, are discussed in this chapter: diabetes, hypertension, cancer, alcoholism, heart disease, hereditary polymorphic light eruption, gallbladder disease, tooth decay, and periodontal disease.

This chapter also covers biochemical or enzymatic variations that are of some health concern: lactose enzyme deficiency, several of the malaria-connected blood and enzyme types, and differential drug metabolism.

Morbidity and mortality patterns differ by racial group. For instance, in 1979 Whites in the United States had an age-adjusted mortality rate of 563 per 100,000 population (National Center for Health Statistics, 1979). This was 33% lower than the Black mortality rate (840 per 100,000). Comparable mortality statistics are not directly available for American Indians and Orientals, since the total "non-White" figures include them with Blacks. However, the total non-White mortality rate is 776 per 100,000, 27% higher than the White rate and 8% lower than for Blacks alone, giving some evidence that United States Orientals and American Indians together have more favorable mortality rates than Blacks.

The diseases that contribute most heavily to these mortality rates are heart disease, cancer, and cerebrovascular diseases. White mortality rates are 24% higher than Black rates for heart disease and 11% higher for cancer. They are similar for cerebrovascular disease (National Center for Health Statistics, 1979).

The 1964 National Health Examination Survey findings showed that White and Black morbidity rates were similar for coronary heart disease and rheumatoid arthritis. Blacks had significantly higher rates of hypertension and hypertensive heart disease, less tooth decay, and more periodontal disease than Whites (Damon, 1971).

Health workers must be cautious when interpreting disease pattern differences between racial groups; the variations may be due to factors other than genetics. Socioeconomic and demographic variables contribute much to what may appear to be racial differences. Certain groups may be younger, poorer, or have less access to health care. Diet, alcohol or tobacco use, and exercise are environmental variables that may make a difference. To demonstrate an actual racial difference is difficult; the other variables must be considered first. Consequently, some differences now thought to be racial may turn out to be caused by something else.

The biochemical variations and diseases covered in this chapter are arranged into three categories. The first consists of biochemical variants and their health effects. Alcohol metabolism, lactose intolerance, conditions associated with resistance to malaria (sickle cell disease, glucose-6-phosphate dehydrogenase deficiency, thalassemia, and Duffy blood grop), and specific drug metabolism differences are in this category.

Diseases of overabundance are the second category. These diseases are relative newcomes to the list of human ailments. They were not present in earlier times, when people ate less and exercised more. Diabetes, various heart diseases, gallbladder disease, tooth decay, and periodontal disease are members of this group.

The third category is a miscellany of two diseases. One is a discrete entity—hereditary polymorphic light eruption—that occurs in American Indians. The other is a diverse group of cancers that are worldwide in their distribution.

BIOCHEMICAL VARIANTS

As researchers learn more about the diseases that afflict human beings, more conditions will be added to this category. At present lactose intolerance and several drug metabolism differences have been well studied. Alcohol metabolism and the malaria-connected conditions are subjects of much current research, with more work needed.

Lactose Intolerance

The inability to digest milk sugar (lactose) is one of the better known conditions that correlate with race. Lactose-intolerant individuals suffer from gastrointestinal symptoms when they ingest milk. The symptoms are dose-dependent—the more milk ingested, the more severe the symptoms (Bayless et al, 1975). Consequently some lactose-intolerant individuals can ingest small quantities of milk with impunity.

Bloating, flatulence, and cramps are the common symptoms associated with lactose intolerance. Diarrhea occurs less commonly; it is associated with ingestion of large amounts of milk—something a lactose-intolerant individual usually avoids. Symptoms occur because lactose is not metabolized. Like any sugar, lactose is subject to bacterial action that produces gas. Sugar has a relatively high osmotic action, attracting water from the intestinal wall, which adds the complications of diarrhea and increased intestinal motility. Symptoms are easily alleviated by avoiding ingestion of milk or milk products. Lactose-intolerant individuals should read food labels, since many packaged foods—processed meats, cake mixes, canned soups, instant potatoes, salad dressings, canned and frozen fruits—have lactose added in processing. Aged cheeses and soured milk products are safe to use, because their lactose has been converted to lactic acid.

The mode of inheritance of lactose intolerance is still not settled. It does run in families (Newcomer et al, 1977; Johnson, Cole, Ahern, 1981), but the specific gene or genes involved in its transmission have not been completely identified (Johnson, Cole, Ahern, 1981; Ho, Povey, Swallow, 1982). The confusion about its inheritance is caused by variability in the time of appearance of symptoms and the severity of symptoms. There seem to be two periods when symptoms of lactose intolerance are likely to be first noted—shortly after weaning and in the teens or early 20s (Simoons, 1969; Lebenthal, Antonowicz, Schwachman, 1975; Caskey et al, 1977). Lactose-intolerant individuals also differ in their ability to consume various amounts of milk. Some can consume a good deal without suffering many symptoms, while others get symptoms from small amounts. These puzzling variations have hindered the genetic search but need not affect health care advice to lactose-intolerant persons. Those who cannot tolerate milk or milk products should avoid milk or consume aged cheeses and sour milk products. For advice on lactose-intolerant children, see the growth and development section of chapter 3.

Lactose intolerance occurs in a majority of the world's population. It affects 94% of Orientals, 90% of African Blacks, 79% of American Indians, 75% of American Blacks, 50% of Mexican-Americans, and 17% of American Whites (McCracken, 1971; Leichter, Lee, 1971; Dill et al, 1972; Duncan, Scott, 1972; Bose, Welsh, 1973; Sowers, Winterfeldt, 1975). This large number of lactose-intolerant people makes American dietary advice concerning the importance of milk and milk products seem narrow and ethnocentric. The approach to nutrition counseling that uses the major dietary staple would avoid this problem. See the section on diet in chapter 5.

Differential Drug Metabolism

There are three classes of reactions to foreign chemicals in the human body: hydrolyses, conjugations, and oxidations (Kalow, 1982). Racial variation occurs in all three reactions. Succinylcholine is a well-studied hydrolyzed drug. Isoniazid, phenobarbital, and amobarbital are examples of conjugation reactions. Primaquine and diphenhydramine (Benadryl) are catalyzed by oxidation. All these drugs are covered in this section.

The enzyme pseudocholinesterase inactivates succinylcholine, a muscle relaxant used in surgery. The drug is inactivated rapidly by most individuals but some people who have an atypical form of the enzyme suffer prolonged muscular paralysis and an inability to breathe without mechanical help for several hours after administration of the drug. Blacks, Orientals, and American Indians are at very low risk of having an atypical form of pseudocholinesterase. Whites are at a slightly higher risk (1 out of 2,500). A few smaller groups—Ashkenazic Jews and Alaskan Eskimos—are at considerably greater risk (Vesell, 1972; Kalow, 1972; Scott, 1973): 1 out of every 135 southwestern Alaskan Eskimos is unable to metabolize succinylcholine normally. For populations at risk for this enzyme defect, a test for the enzyme deficiency shold be done before surgery (Motulsky, 1970).

People react in one of two ways to the metabolism of isoniazid, a drug used in tuberculosis therapy. They inactivate the drug either rapidly or slowly (Evans, Manley, McKusick, 1960). Those who inactivate it slowly are more prone to developing peripheral neuropathy during isoniazid therapy (Vesell, 1972). They are also prone to toxic reactions from dapsone, hydralazine, phenelzine, sulfamaprine, and sulfamethazine (Vesell, 1972). Because of the possibility of neuropathy, pyridoxine is routinely given with isoniazid. During the treatment of tuberculosis, when isoniazid doses are spaced at long intervals, slow inactivators respond to therapy more satisfactorily than rapid inactivators (LaDu, 1972). Rapid inactivation of isoniazid occurs in 40% of Whites, 60% of Blacks, 60–95% of American Indians and Eskimos (depending on the tribe), and 85–90% of Orientals (LaDu, 1972; Hodgkin, Eidus, Biley, 1979).

Recent work indicates some differences in barbiturate metabolism. Orientals differ from Whites in metabolizing this drug group (Kalow, 1982). What this means clinically is yet to be determined.

Glucose-6-phosphate dehydrogenase (G6PD) deficiency is also discussed in the section on conditions associated with malaria. This enzyme deficiency, which is fairly common in people originating from areas where malaria is common, produces hemolytic anemia when certain drugs are administered. The following drugs may cause severe hemolytic anemia in such persons: analgesics (e.g., aspirin, phenacetin), sulfonamides and sulfones, antimalarials (e.g., primaquine, atabrine), antibacterials (e.g., nitrofurantoin, chloramphenicol, para-aminosalicylic acid), and vitamin K, probenecid, and quinidine. Over 30 drugs are known to cause the problem in G6PD-deficient individuals (Vesell, 1972).

Knowledge of differential drug metabolism is beginning to increase. One physician noted differences in reaction to diphenhydramine (Benadryl) between his Oriental and his White patients. This clinical observation led to a study showing that the Orientals had half the plasma levels of the drug that White subjects had (Spector et al, 1980). They also experienced less sedation effect and less deterioration in psychomotor tasks. The authors suggested that Oriental adults require larger doses, per body weight, to achieve comparable results.

Alcohol Metabolism

Behavioral scientists have been concerned with alcoholism for many years. In the early 1970s scientists started searching for biologic reasons why Orientals and American Indians reacted differently to alcohol, compared with Blacks and Whites. Many Orientals and American Indians experience marked facial flushing and other vasomotor symptoms after drinking alcohol. Most Whites do not experience these symptoms (Wolff, 1973). The incidence of alcoholism is higher among American Indians than among Blacks and Whites, while Orientals seem to have a low alcoholism rate (Fenna et al, 1971; Klatsky et al, 1977).

These observations led to the discovery of differences in alcohol metabolism and to the description of two enzymes involved in alcohol metabolism. Alcohol is first oxidized to acetaldehyde by the alcohol dehydrogenase (ADH) enzyme; then acetaldehyde is oxidized to acetic acid by acetaldehyde dehydrogenase (ALDH) (Stamatoyannopoulos, Chen, Fukui, 1975). There are two variants of the ADH enzyme. The high-activity type converts alcohol to acetaldehyde quickly, while the low-activity variant converts it slowly (Kalow, 1982). The ALDH enzyme also comes in several varieties: $ALDH_1$, through $ALDH_4$. $ALDH_1$ is considered "normal," the other three types "deficient" in the ability to oxidize acetaldehyde (Goedde et al, 1983). The oxidation of acetaldehyde to acetic acid is delayed in people deficient in ALDH enzyme.

Individuals who have the high-activity ADH variant and are also "deficient" in ALDH enzyme oxidize alcohol to acetaldehyde quickly and then experience a long period of high blood acetaldehyde levels because the acetaldehyde breaks down slowly. It is thought that high acetaldehyde levels produce many of the symptoms of alcohol intoxication, including facial flushing and other vasomotor symptoms (Mizoi et al, 1979). Consequently, the interactions of these alcohol metabolism enzymes may explain differences in reaction to alcohol ingestion.

Studies show that 85–90% of American Indians and Orientals have the high-activity variant of ADH. This means the alcohol they drink is rapidly changed to acetaldehyde. In contrast, most Blacks and Whites (90–100%) have the low-activity variant; they slowly change alcohol to acetaldehyde (Murray, Price, 1972; Goedde et al, 1983). The next step, conversion of acetaldehyde to acetic acid, is delayed in 35–70% of American Indians and Orientals, whereas it proceeds at the faster rate in most Blacks and Whites (Goedde et al, 1983). Because of these enzyme differences, a majority of Orientals and American Indians experience a rapid onset but a slow decrease of blood acetaldehyde levels. This leaves them with a long period of exposure to the substance that is thought to cause many of the symptoms of alcohol intoxication.

While the two alcohol metabolism enzymes and their variants help explain the differences in reaction to alcohol ingestion, they do not help explain the different rates of alcoholism between Orientals and American Indians.

Malaria-connected Conditions

Malaria has brought about the evolutionary development of several hemoglobin and enzyme variants. These variants spread in the tropical and subtropical parts of the world in response to the severe stress of malarial disease. While providing some protection from malaria, they also produce less advantageous effects. Four of the more common malaria-specific variants are sickle cell hemoglobin, thalassemia, glucose-6-phosphate dehydrogenase (G6PD) deficiency, and Duffy blood group.

Sickle cell hemoglobin Sickle cell hemoglobin is the best known condition related to malaria. Sickle cell homozygotes, who have received the sickle gene from both parents, usually die during childhood. Normal homozygotes, who have two normal hemoglobin genes, get full-blown cases of falciparum malaria when they are infected. Heterozygotes, with one normal and one sickle cell gene, have less severe cases of falciparum malaria. It is thought that, as the malaria parasite attempts to reproduce in heterozygous individuals, cells with the sickle defect are destroyed before the parasite's reproduction is complete. This reduces the parasite load. These infected cells sickle and are then eliminated from circulation by phagocytosis (Roth et al, 1978).

Blacks have the highest incidence of sickle cell gene of any group—about 9.5% of American Blacks and up to 20% of African Blacks (Maugh, 1981). American Blacks have a lower incidence of the gene than African blacks partly because of the admixture of White genes and partly because there is no longer an advantage for the heterozygote in the United States, since malaria has been virtually eliminated. Although Blacks have a much higher incidence of sickle cell disease than other races, individuals of Greek, Italian, Turkish, and Arabian ancestry occasionally have cases (Campbell, Oski, 1977). Sickle cell trait was found in 0.08% of United States naval recruits. Individuals with the trait were of Germanic, Italian, and American Indian origin (McGrew, 1973). There is no need to postulate Black ancestors to explain the presence of sickle cell trait.

There seem to be two levels of severity of the disease. The less severe form is present more often in the non-Black population, while the more severe, classical form is most prevalent in Blacks (Campbell, Oski, 1977; Dover, Boyer, Pembrey, 1981), although Blacks can have either form (Dover, Boyer, Pembrey, 1981). The presence of the less severe form in non-Black populations and the relative rarity of the disease in these populations have probably led to the misconception that sickle cell disease is exclusively a disease of Blacks.

Thalassemia Thalassemia is a genetic condition that results in various degrees of anemia, depending on the particular variety of the disorder. The clinical symptoms result from a defect in the rate of synthesis of normal hemoglobin rather than from production of an abnormal variant. This very complex syndrome varies depending on whether the defect is on the alpha or beta chain of the hemoglobin molecule, and which type of defect occurs on the chain (Bodmer, Cavalli-Sforza, 1976).

Heterozygotes for any of the thalassemia varieties have mild to absent clinical signs of anemia, but homozygotes die of anemia, because there is no cure. Like sickle cell hemoglobin, thalassemia is prevalent in malarial areas, and it is thought to protect heterozygotes from severe malaria illness.

The incidence of thalassemia is particularly high in people from the Mediterranean region, the Middle East, and Southeast Asia. Italy has a 7–15% incidence (Marinucci et al, 1982). Greece, India, Sudan, Turkey, Israel, and

West Africa all have high frequencies, ranging from 0% to 50% of the people in a particular area (Bodmer, Cavalli-Sforza, 1976; Weatherall, Clegg, 1982). The condition is rare in the Japanese (Shibata, Miyaji, Ohba, 1980).

Glucose-6-phosphate dehydrogenase (G6PD) deficiency Another enzyme defect, G6PD deficiency, also causes anemia under certain conditions and is common in malarial areas. Approximately 100 million people in the world are affected. There are four main G6PD blood types—normal A and B, and A− and B− (Mediterranean)—but over 80 variants (Vesell, 1972). The A− and B− types are deficient in the G6PD enzyme. The malaria parasite prefers to attack red blood cells that contain the G6PD enzyme. Cells with the deficiency are less likely to be parasitized. This lowers the parasite load, lessening the severity of the illness.

G6PD is X-linked; therefore, problems related to the deficiency occur mostly in males. The deficiency occurs in Blacks, Orientals, and people from Mediterranean areas. Blacks usually have the A− variety, while Orientals and Mediterranean people generally have the B− variety. Individuals with B− deficiency experience hemolysis after eating fava beans. Both types of deficiency cause hemolysis after receiving certain drugs. See the discussion of differential drug metabolism later in this chapter for information on specific drugs.

Duffy blood group A particular blood type, Duffy negative, protects individuals against vivax malaria without "costing" anything. It does not cause death in homozygotes, nor does it produce anemia. Blacks are almost exclusively Duffy negative blood types, while the rest of the world's populations are mostly Duffy positive. American Blacks do have some Duffy positive because of their admixture with Whites.

It is postulated that the Duffy positive red blood cells have receptors on their surfaces for the vivax malaria parasites, while Duffy negative cells lack these receptors (Bodmer, Cavalli-Sforza, 1976). This is supported by evidence that, when individuals are exposed to the parasite under the same conditions, those who are Duffy negative do not get vivax malaria, while Duffy positive individuals do (Spencer et al, 1978).

DISEASES OF OVERABUNDANCE

The diseases of overabundance are generally preventable. Some, such as heart disease, cerebrovascular disease, and diabetes, have declined in incidence over the past 30 years (National Center for Health Statistics, 1979). The declining diseases may be responding to increasing awareness and use of improved diets and more exercise, as well as to earlier diagnosis and better treatment. This section will also consider gallbladder disease, tooth decay, and periodontal disease, all diseases of abundance.

Diabetes

Diabetes is a well-studied but still poorly understood disease. Diabetes rates differ among groups, but the differences seem due not as much to race as to the degree and duration of sedentary living by the population. When sedentary living becomes the norm, obesity increases, and diabetes and obesity are closely entwined health problems. They occur together and in high frequency in many American Indian tribes and Pacific Islanders (West, 1978; Prior et al, 1978). There are also pockets of high diabetes and obesity frequency in other areas of the world—among Japanese in Hawaii and Indians in South Africa (Kawate et al, 1978; Gupte, Joshi, Dave, 1978). The risk of developing diabetes has been calculated for Pima Indians based on the degree of obesity (Hammon, Bennett, Miller, 1978). Individuals in the heaviest quartile have almost three times the risk of developing diabetes compared to the slimmest quartile. On counseling groups at risk, see chapters 5 and 8.

It has been hypothesized that a diabetic tendency is an advantage in situations where people experience famine cycles (Neel, 1963). Individuals with a disposition toward diabetes tend to put on weight easily when food is plentiful. The stored fat comes in handy during a famine. Such people have a buffer against environmental stress, can reproduce more efficiently, and hence can pass on their diabetic tendencies to the next generation.

Although this hypothesis neatly explains the high incidence in certain groups, it does not explain the low incidence in other, more stable groups. It seems to me that the more stable groups went through the same process, but many generations ago, when there was no adequate treatment for diabetes. Consequently, large numbers of people with diabetes were eliminated from the population due to decreased fertility and excess mortality, and genetic factors conducive to diabetes became less common. If this is so, the groups with recently changed food habits may continue to have an excess burden of diabetes compared to the groups that experienced these changes earlier in history. This could mean that many individuals from high-risk groups constantly need to monitor their weight and the weight of their children. Perhaps if such groups remain together as communities, a cultural response to the shared problem may develop.

Heart Disease and Hypertension

Diseases of the heart are the leading cause of death in the United States. This category includes coronary artery and hypertensive heart disease, both of which vary in frequency by racial group.

Coronary artery disease White and Blacks have similar age-adjusted rates of coronary artery disease mortality (Weisse, Abiuso, Thind, 1977; Gordon, 1976). Few data are available for comparisons with other groups, but a Hawaiian study indicates that White mortality rates are higher than those of either Japanese-Americans or Filipino-Americans (Gerber, 1980). This study also showed that coronary artery disease mortality is higher in urban than in rural areas.

The mortality difference among races is probably not due to genetic factors. The Hawaiian study demonstrated that mortality rose in people from a given group (Japanese or Filipino) when they moved from rural to urban areas (Gerber, 1980). With a move to the city come changes in living style—in particular, changes in diet and work patterns. Research has convincingly demonstrated a connection between a high rate of coronary artery disease and sedentary living patterns. One effect of a sedentary lifestyle is to increase total cholesterol blood levels, specifically low density lipoproteins. Such a condition is conducive to development of coronary artery disease. Physical exercise lowers total cholesterol levels and causes the body to produce more high density lipoproteins, which seem to have a protective effect against coronary artery disease (Clarkson et al, 1981; Hartung et al, 1980). Sedentary people need to add a regular routine of physical exercise to their life-styles, to replace what they would have had to do if they were still providing their own food and shelter from scratch.

Hypertension The rates of hypertension and hypertensive heart disease are high in Blacks compared to other groups, but the cause is not clear (Gordon, 1976; Rowland, Roberts, 1982). More and more evidence suggests that environment is important in determining this difference. This is borne out by the rising rates of hypertension in groups such as American Indians and South Pacific Island natives (Strotz, Shorr, 1973; Sievers, 1977; DeStefano, Coulehan, Wiant, 1979; Prior, 1971) and the decreasing rates in others, such as Blacks and Whites (Kopstein, 1980; Metropolitan Life Insurance Company, 1980).

Psychologic stress, low socioeconomic status, urbanization, a decrease in physical activity, and overweight have all been implicated in the causation of hypertension. (See section on aging in chapter 3 for more information on environmental variables that promote hypertension.) Besides these variables, some data suggest that Blacks may have a genetic predisposition to the disease. Compared to Whites, Blacks get hypertension at younger ages (Roberts, Maurer, 1977), have higher age-standardized rates of stroke, and get first strokes at earlier ages (Moss, 1981). Hypertension seems to be more severe and less responsive to treatment in Blacks than in Whites (Relman, 1982).

Blacks also have more renal pathology related to hypertension than do Whites (Levy et al, 1978; Rostand et al, 1982). Differences in renal response to hypertension start early. A large study of children 5–14 years of age showed that Black children with elevated blood pressures had low renin activity, whereas their White counterparts had high renin activity (Voors et al, 1979). These same White children with high blood pressure also had higher heart rates than their Black counterparts. Blacks also seem to have a more efficient sodium sparing ability, which helps them withstand heat stress; they lose less salt during sweating than whites do (Riggs, Sargent, 1964; Dill et al, 1983). This salt retention may aggravate hypertension.

These observations lend support to the notion of a genetic difference in blood pressure regulation between Blacks and Whites. This possibility should be considered in hypertension prevention and treatment programs. Low-salt diets may be more effective in treating Black hypertensives, and medication adjustments may be needed for different racial groups. Research is warranted in these two areas.

Gallbladder Disease

Physicians who work with American Indians have noted a high frequency of gallbladder disease in this group. They mention that cholecystectomy is the Indians' most commonly performed surgical procedure (Brown, Christensen, 1967; Comess, Bennett, Burch, 1967). The magnitude of the American Indians' high frequency of gallbladder disease is illustrated by a study comparing Pima Indians and Massachusetts Whites (Comess, Bennett, Burch, 1967): 36% of the Indians had hospital admissions for gallbladder disease compared to 6% of the Whites. Other studies have corroborated these findings (Brown, Christensen, 1967; Erickson, 1972; Boss et al, 1982). Data on other races are limited, but a Hawaiian study examined hospital admission rates for gallbladder disease in eight races—Chinese, Japanese, Koreans, Filipinos, Hawaiians, Portuguese, Puerto Ricans, and Whites (Yamase, McNamara, 1972). While the differences in admission rates among the races were statistically significant, they were hardly clinically significant, since the rates varied between only 1% and 2.5%. The Hawaiian gallbladder admission rates were a few percentage points lower than those of the Massachusetts Whites but were over 30 percentage points lower than those of the Pima Indians.

The sex ratio for gallbladder disease sufferers is around two females to one male for most groups. American Indians have a very different sex ratio: approximately seven females to one male (Yamase, McNamara, 1972).

Several variables have been compared in individuals with and without gallbladder disease; only pregnancy is associated (Comess, Bennett, Burch,

1967). The mean number of pregnancies is higher in women with gallbladder disease. This holds for all the races studied. Gallbladder disease is not associated with diabetes, obesity, or serum cholesterol level (Comess, Bennett, Burch, 1967). Not enough data are available to determine whether the American Indians' high rate of gallbladder disease is genetic or environmental.

Tooth Decay and Periodontal Disease

Whites have more tooth decay and less periodontal disease than Blacks (Johnson, Kelly, VanKirk, 1965; Kelly, Harvey, 1979), and these differences are present from childhood through old age. Complete loss of teeth occurs more often in Whites than in Blacks, despite the higher periodontal disease rate in Blacks (Kelly, VanKirk, Garst, 1967; Burnham, 1974). Some 33% of Whites 45 years of age and older have lost all their teeth, compared with 25% of Blacks in the same age groups (Kelly, VanKirk, Garst, 1967).

The differences in tooth decay between the two races can be explained by assuming that Blacks have harder or denser tooth enamel, which is similar to their denser bones, and further assuming that the harder tooth enamel is less susceptible to the organisms that cause tooth decay. Periodontal disease differences may also be explained: Blacks have poorer oral hygiene than Whites at all ages (Kelly, Sanchez, 1972; Sanchez, 1975; Johnson, Kelly, VanKirk, 1965; Kelly, Harvey, 1979), and poor oral hygiene is related to periodontal disease (Kelly, VanKirk, Garst, 1966).

Improved oral hygiene would probably reduce periodontal disease, and tooth loss in Blacks. Then the superiority of Blacks in overall dental health would be even more pronounced.

MISCELLANEOUS DISEASES

Of the two remaining diseases discussed in this section, hereditary polymorphic light eruption shows a distinct racial distribution that is genetically caused. While cancers occur in some racial groups more than others, evidence is still lacking on the parts that genetic and environmental variables play in the occurrence of various types of cancer.

Hereditary Polymorphic Light Eruption

Hereditary polymorphic light eruption (HPLE) is a skin disease, described in many North and Southern American Indian tribes, that results from exposure to sunlight (Schenck, 1960; Everett et al, 1961; Birt, 1968; Fusaro,

Johnson, 1980). The disease is considered an autosomal dominant condition that occurs in around 1% of some Native American groups (Birt, 1968; Fusaro, Johnson, 1980).

HPLE generally begins during childhood in affected individuals; 35% show symptoms by 5 years of age, 75% by age 16 (Birt, Davis, 1975). The lesions are characterized by erythema, edema, and vesicles with exudation. They often become secondarily infected, resulting in impetigo. Because of a streptococcal affinity for this type of lesion, glomerulonephritis is a sequela of some of the secondary infections. The affected skin areas are always parts of the body that get frequent sun exposure—most often the face and neck, followed by the lower arms and hands (Everett et al, 1961). Parts of the body covered by clothing or protected by hair are free of the disease. Foreheads protected by bangs, necks protected by long hair, under-chins shaded by the head, and upper arms and body protected by clothing are disease-free. This distribution is one of the disease's diagnostic characteristics.

Generally a family history of the condition and a clearing of the skin during the winter are additional clues to its diagnosis. Because HPLE is unique to American Indians, it is probably misdiagnosed frequently. Those who work with Native American groups need to monitor the dermatologic complaints of their patients to determine whether this preventable and treatable condition might be the cause of their problems.

The skin of Indians with HPLE is sensitive to the long ultraviolet rays of the sun; this makes sunscreens ineffective (Fusaro, Johnson, 1980). The skin must be protected by wearing a wide-brimmed hat, high collar, and long sleeves. A few systemic drugs also seem to help (Birt, Davis, 1975; Fusaro, Johnson, 1980).

Cancer

Several cancers appear to vary racially in their incidence, but a closer look often reveals diet or some other environmental factor to be a better predictor of the cancer rate than race. Examples of cancers associated with environmental variables that mimic racial variation will be mentioned in this section, along with a few cancers that show racial preferences where an associated environmental variable is yet to be discovered.

Breast cancer The best evidence for a racial difference in cancer rates exists for breast cancer in women. In the United States, White women have a higher rate of breast cancer than Black women, 75 per 100,000 versus 58 per 100,000 (Asire, Shambaugh, 1976). But Blacks have a lower survival rate after diagnosis than Whites. The differences in both the rates of disease and the survival rates are thought to be due to socioeconomic factors. In the past several years, the Black survival rate has improved (Asire, Shambaugh, 1976).

Comparable incidence rates for Asian Americans and Native Americans are not available, but death rates from breast cancer can be compared among the four groups. Asian Americans and American Indians experience much lower mortality from breast cancer than either Blacks or Whites (Buell, Dunn, 1965; Petrakis, 1971). The search for causes of this difference has led to the discovery of an association between earwax type and breast cancer incidence. Women with dry earwax have a lower incidence of breast cancer than those with wet earwax (Petrakis, 1971). Most Orientals and Native Americans have dry earwax, while most Blacks and Whites have wet earwax. The association of earwax type and breast cancer makes sense histologically, since the glands in the ear canal and the breast are both apocrine glands and their secretions are similar (Petrakis, 1971). Breast cancer rates in Japanese women with wet earwax are twice those of Japanese women with dry earwax (Petrakis, 1971).

The fact that Asian migrants to this country experience a rise in breast cancer rates, but not to the level of the United States residents, indicates that breast cancer causation is complex (Buell, Dunn, 1965; Haenszel, Kurihara, 1968). Diet and age of bearing the first child are also associated with breast cancer, and both variables usually change after migration (Doll, Peto, 1981).

Cancer of the prostate While prostatic cancer rates differ dramatically between Blacks and Whites, the differences are definitely not racial. Formerly Blacks had lower rates than Whites; now their rates are higher (Ernster et al, 1978). A meticulous investigation into the annual age-specific mortality rates for prostatic cancer in Blacks has demonstrated that these rates peaked in men born between 1896 and 1900. Men born before these dates had steadily increasing rates and men born after these dates had slowly declining rates of prostatic cancer. No such pattern was seen in White men. The authors suggest that some variable that became more frequent previous to the 1896–1900 time period and less frequent thereafter has produced the pattern. The authors do not speculate on what the particular variable might be.

Skin cancer Skin cancers of all types are less common in Blacks than in Whites (Fleming et al, 1975). The reason for this is dark skin color and the protection it gives. There are some differences between Blacks and Whites in the types and locations of their various skin cancers. For both, the most common skin cancer is epidermoid carcinoma. In Whites these cancers are generally located on sun-exposed areas of the skin, while in Blacks they more often develop in burn scars and chronic infection sites (Fleming et al, 1975). Squamous cell carcinoma has a higher incidence of regional node metastasis in Blacks compared to Whites. But, when the metastasis rate is compared by site of initial lesion, Whites and Blacks whose cancer is unrelated to sun exposure have similar rates (Fleming et al, 1975). Evidently sun-caused cancer is easier to detect at an earlier stage of development and is thus more amenable to cure.

For Whites the most commonly treated serious skin cancer is basal cell carcinoma, while in Blacks this type is relatively rare (Fleming et al, 1975). The fact that this cancer occurs on sun-exposed areas of the skin in both Blacks and Whites accounts for its rarity in Blacks, who are protected to a greater extent by their dark skin.

Melanoma is not a common cancer in either Blacks or Whites, but when it does occur it is over 100 times more frequent in Whites (Krementz et al, 1976). Another striking difference is that Blacks predominantly develop melanomas on the palmar surface of the hands or plantar surface of the feet; over 50% of their lesions occur at these two sites (Krementz et al, 1976). In Whites, lesions occur most often on sun-exposed surfaces—the head and neck in men and the lower extremities in women (Luce, McBride, Frei, 1974). Survival after diagnosis is generally poor, but when the lesion is discovered in stage I (the early stage), survival is similar and good for both races (Krementz et al, 1976).

Digestive tract cancers Cancers of the esophagus, stomach, and colon show distinct geographic areas of high and low frequency. These areas generally involve specific racial groups, but dietary items rather than genetic factors are mainly implicated in digestive cancer causation. Recent research has generally determined the ingested substances that are cancer-promoting. High-fat and high-meat diets are probably the major promoters of digestive cancer (Armstrong, Doll, 1975; Wynder, 1979; Doll, Peto, 1981). Smoking is a close contender for next place, especially when combined with high alcohol consumption. Several vegetables seem to protect against cancer (Doll, Peto, 1981). The dietary evidence is now so strong that racial variation is not seriously considered.

Summary

Several present-day physical afflictions have, sometime in the past, been helpful body responses (lactose intolerance, malaria-connected conditions, and diabetes). Some of the biologic variations noted in this chapter are specifically racial (alcohol metabolism, lactose intolerance, malaria-connected conditions, drug metabolism differences, tooth decay, hereditary polymorphic light eruption, and perhaps gallbladder disease). Several of the variations occur more often in one group than another, but environmental variables are more important causes than racial variation (diabetes, heart diseases, periodontal disease, and cancer).

REFERENCES

Armstrong B, Doll R: Environmental factors and cancer incidence and mortality in different countries, with special reference to dietary practices. *Int J Cancer* 1975; 15:617–631.

Asire A, Shambaugh E: Cancer of the breast. Pages 157–163, in: *Cancer Patient Survival.* Axtell L, Asire A, Myers M (editors). US Department of Health, Education, and Welfare, Bethesda, MD, 1976.

Bayless T et al: Lactose and milk intolerance: Clinical implications. *N Engl J Med* 1975; 292(22):1156–1159.

Birt A: Photodermatitis in Indians of Manitoba. *Can Med Assoc J* 1968; 98:392–397.

Birt A, Davis R: Hereditary polymorphic light eruption of American Indians. *Int J Dermatol* 1975; 14:105–111.

Bodmer W, Cavalli-Sforza L: *Genetics, Evolution, and Man*. Freeman, 1976.

Bose D, Welsh J: Lactose malabsorption in Oklahoma Indians. *Am J Clin Nutr* 1973; 26:1320–1322.

Boss L et al: Cancers of the gallbladder and biliary tract in Alaskan natives: 1970–79. *J Natl Cancer Inst* 1982; 69(5):1005–1007.

Brown J, Christensen C: Biliary tract disease among the Navajos. *JAMA* 1967; 202(11):138–139.

Buell P, Dunn J: Cancer mortality among Japanese Issei and Nisei of California. *Cancer* 1965; 18(5):656-664.

Burnham C: Edentulous persons. *Vital Health Stat* 1974; 10(89):1–31.

Campbell J, Oski F: Sickle cell anemia in an American White boy of Greek ancestry. *Am J Dis Child* 1977; 131:186–188.

Caskey D et al: Effects of age on lactose malabsorption in Oklahoma Native Americans as determined by breath H_2 analysis. *Am J Dig Dis* 1977; 22(2):113–116.

Clarkson P et al: High density lipoprotein cholesterol in young adult weight lifters, runners, and untrained subjects. *Hum Biol* 1981; 53(2):251–257.

Comess L, Bennett P, Burch T: Clinical gallbladder disease in Pima Indians: Its high incidence in contrast to Framingham, Massachusetts. *N Engl J Med* 1967; 277(17):894–898.

Damon A: Race, ethnic group, and disease. Pages 57-74 in: *The Biological and Social Meaning of Race*. Osborne R (editor). Freeman, 1971.

DeStefano F, Coulehan J, Wiant M: Blood pressure survey on the Navajo Indian reservation. *Am J Epidemiol* 1979; 109(3):335–345.

Dill J et al: Lactase deficiency in Mexican-American males. *Am J Clin Nutr* 1972; 25:869–870.

Dill D et al: Volume and composition of hand sweat of White and Black men and women in desert walks. *Am J Phys Anthropol* 1983; 61(1):67–73.

Doll R, Peto R: Avoidable risks of cancer in the U.S. *J Natl Cancer Inst* 1981; 66(6):1196–1256.

Dover G, Boyer S, Pembrey M: F-cell production in sickle cell anemia: Regulation by genes linked to β-hemoglobin locus. *Science* 1981; 211:1441–1443.

Duncan I, Scott E: Lactose intolerance in Alaskan Indians and Eskimos. *Am J Clin Nutr* 1972; 25:867–868.

Erickson W: Cholecystectomy and the American Indian. *Rocky Mt Med J* (July) 1972; 69(7):33–35.

Ernster V et al: Prostatic cancer: Mortality and incidence rates by race and social class. *Am J Epidemiol* 1978; 107(4):311–319.

Evans D, Manley K, McKusick V: Genetic control of isoniazid metabolism in man. *Br Med J* (August) 1960; 13:485–491.

Everett M et al: Light-sensitive eruptions in American Indians. *Arch Dermatol* 1961; 83:105–110.

Fenna D et al: Ethanol metabolism in various racial groups. *Can Med Assoc J* 1971; 105:472–475.

Fleming I et al: Skin cancer in Black patients. *Cancer* 1975; 35:600–605.

Fusaro R, Johnson J: Hereditary polymorphic light eruption in American Indians: Photoprotection and prevention of streptococcal pyoderma and glomerulonephritis. *JAMA* 1980; 244(13):1456–1459.

Gerber L: The influence of environmental factors on mortality from coronary heart disease among Filipinos in Hawaii. *Hum Biol* 1980; 52(2):269–278.

Goedde H et al: Population genetic studies on aldehyde dehydrogenase isozyme deficiency and alcohol sensitivity. *Am J Hum Genet* 1983; 35:769–772.

Gordon T: Heart disease in adults. *Vital Health Stat* 1976; 11(6):1–43.

Gupte O, Joshi M, Dave S: Prevalence of diabetes in India. *Adv Metab Disord* 1978; 9:147–166.

Haenszel W, Kurihara M: Studies of Japanese migrants: I. Mortality from cancer and other diseases among Japanese in the United States. *J Natl Cancer Inst* 1968; 40(1):43–68.

Hammon R, Bennett P, Miller M: Incidence of diabetes among the Pima Indians. *Adv Metab Disord* 1978; 9:49–63.

Hartung G et al: Relation of diet to high-density-lipoprotein cholesterol in middle-aged marathon runners, joggers, and inactive men. *N Engl J Med* 1980; 302(7):357–361.

Ho M, Povey S, Swallow D: Lactase polymorphism in adult British natives: Estimating allele frequencies by enzyme assays in autopsy samples. *Am J Hum Genet* 1982; 34:650–657.

Hodgkin M, Eidus L, Biley W: Isoniazid phenotyping of Black as well as White parents. *Can J Physiol Pharmacol* 1979; 57:760–763.

Johnson E, Kelly J, VanKirk L: Selected dental findings for adults. *Vital Health Stat* 1965; 11(7):1–23.

Johnson R, Cole R, Ahern F: Genetic interpretation of racial/ethnic differences in lactose absorption and tolerance. *Hum Biol* 1981; 53(1):1–13.

Kalow W: The metabolism of xenobiotics in different populations. *Can J Physiol Pharmacol* 1982; 60:1–9.

Kalow W: Pharmacogenetics of drugs used in anaesthesia. Pages 415–427 in: *Human Genetics*. Proc 8th Intl Cong Hum Genet. Excerpta Medica, 1972.

Kawate R et al: Preliminary studies of the prevalence and mortality of *diabetes mellitus* in Japanese in Japan and on the island of Hawaii. *Adv Metab Disord* 1978; 9:202–224.

Kelly J, Harvey C: Basic dental examination findings of persons 1–74 years. *Vital Health Stat* 1979; 11(214):1–33.

Kelly J, Sanchez M: Periodontal disease and oral hygiene among children. *Vital Health Stat* 1972; 11(117):1–28.

Kelly J, VanKirk L, Garst C: Oral hygiene in adults. *Vital Health Stat* 1966; 11(16):1–30.

Kelly J, VanKirk L, Garst C: Total loss of teeth in adults. *Vital Health Stat* 1967; 11(27):1–23.

Klatsky A et al: Alcohol consumption among White, Black, or Oriental men and women: Kaiser-Permanente multiphasic health examination data. *Am J Epidemiol* 1977; 105(4):311–322.

Kopstein A: Hypertension-related mortality by health service area 1968–72. *Natl Center Health Stat* 1980; 11:1–115.

Krementz E et al: Malignant melanoma in the American Black. *Ann Surg* 1976; 183(5):533–541.

LaDu B: Isoniazid and pseudocholinesterase polymorphisms. *Fed Proc* 1972; 31(4):1276–1285.

Lebenthal E, Antonowicz I, Schwachman H: Correlation of lactase activity, lactose tolerance, and milk consumption in different age groups. *Am J Clin Nutr* 1975; 28:595–600.

Leichter J, Lee M: Lactose intolerance in Canadian west coast Indians. *Am J Dig Dis* 16(9):809–813.

Levy S et al: Renal vasculature in essential hypertension: Racial differences. *Ann Intern Med* 1978; 88(1):12–16.

Luce J, McBride C, Frei E: Melanoma. Pages 1823–1843 in: *Cancer Medicine*. Holland J, Trei E (editors). Lea & Febiger, 1974.

Marinucci M et al: Hemoglobinopathies in Italy. *Hemoglobin* 1982; 6(3):247–255.

Maugh T: A new understanding of sickle cell emerges. *Science* 1981; 211:265–267.

McCracken R: Lactase deficiency: An example of dietary evolution. *Curr Anthropol* 1971; 12(4–5):479–517.

McGrew C: Sickle cell trait in the White population. *JAMA* 1973; 224(13):1762–1763.

Metropolitan Life Insurance Company: Cardiovascular diseases—United States, Canada, and western Europe. *Stat Bull Metropol Life Ins Co* 1980; 61(4):8–12.

Mizoi Y et al: Relationship between facial flushing and blood acetaldehyde levels after alcohol intake. *Pharmacol Biochem Behav* 1979; 10:303–311.

Moss A: Stroke survivors among the noninstitutionalized population 20 years of age and over: United States, 1977. *Advancedata* 1981; 68:1–15.

Motulsky A: General remarks on genetic factors in anesthesia. *Humangenetik* 1970; 9:246–248.

Murray R, Price P: Ontogenetic, polymorphic, and interethnic variation in the isoenzymes of human alcohol dehydrogenase. *Ann NY Acad Sci* 1972; 197:68–72.

National Center for Health Statistics: Advance report of final mortality statistics, 1979. *Monthly Vital Stat Rep* 1979; 31(6-Suppl):1–59.

Neel J: Diabetes mellitus: A "thrifty" genotype rendered detrimental by "progress"? *Am J Hum Genet* 1963; 14(4):353–362.

Newcomer A et al: Family studies of lactase deficiency in the American Indian. *Gastroenterology* 1977; 73(5):985–988.

Petrakis N: Cerumen genetics and breast cancer. *Science* 1971; 173:347–349.

Prior I: The price of civilization. *Nutri Today* (July/August) 1971; 6:2–11.

Prior I et al: The relationships of diabetes, blood lipids, and uric acid levels in Polynesians. *Adv Metab Disord* 1978; 9:242–262.

Relman A: Race and end-stage renal disease. *N Engl J Med* 1982; 306(21):1290–1291.

Riggs S, Sargent F: Physiological regulation in moist heat by young American Negro and White males. *Hum Biol* 1964; 36(4):339–353.

Roberts J, Maurer K: Blood pressure levels of persons 6–74 years. *Vital Health Stat* 1977; 11(203):1–103.

Rostand S et al: Racial differences in the incidence of treatment for end-stage renal disease. *N Engl J Med* 1982; 306(21):1276–1279.

Roth E et al: Sickling rates of human AS red cells infected in vitro with Plasmodium falciparum malaria. *Science* 1978; 202:650–652.

Rowland M, Roberts J: Blood pressure levels and hypertension in persons ages 6–14 years: United States, 1976–80. *Advancedata* 1982; 84:1–10.

Sanchez M: Oral hygiene among youths 12–17 years. *Vital Health Stat* 1975; 11(151):1–36.

Schenck R: Controlled trial of methoxsalen in solar dermatitis of Chippewa Indians. *JAMA* 1960; 172(11):1134–1137.

Scott E: Inheritance of two types of deficiency of human serum cholinesterase. *Ann Hum Genet* 1973; 37:139–143.

Shibata S, Miyaji T, Ohba Y: Abnormal hemoglobins in Japan. *Hemoglobin* 1980 4(3–4): 395–408.

Sievers M: Historical overview of hypertension among American Indians and Alaskan natives. *Ariz Med* 1977; 34(9):607–610.

Simoons F: Primary adult lactose intolerance and the milking habit: A problem in biological and cultural interrelations. *Am J Dig Dis* 1969; 14(12):819–836.

Sowers M, Winterfeldt E: Lactose intolerance among Mexican Americans. *Am J Clin Nutr* 1975; 28:704–705.

Spector R et al: Diphenhydramine in Orientals and Caucasians. *Clin Pharmacol Ther* 1980; 28(2):229–234.

Spencer H et al: The Duffy blood group and resistance to Plasmodium vivax in Honduras. *Am J Trop Med Hyg* 1978; 27(4):664–670.

Stamatoyannopoulos G, Chen S, Fukui M: Liver alcohol dehydrogenase in Japanese: High population frequency of atypical form and its possible role in alcohol sensitivity. *Am J Hum Genet* 1975; 27:789–796.

Strotz C, Shorr G: Hypertension in the Papago Indians. *Circulation* 1973; 48:1299–1303.

Vesell E: Drug therapy—pharmacogenetics. *N Engl J Med* 1972; 287(18): 904–909.

Voors A et al: Racial differences in blood pressure control. *Science* 1979; 204:1091–1094.

Weatherall D, Clegg J: Thalassemia revisited. *Cell* 1982; 29:7–9.

Weisse A, Abiuso P, Thind I: Acute myocardial infarction in Newark, N.J.—a study of racial incidence. *Arch Intern Med* 1977; 137:1402–1405.

West K: Diabetes in American Indians. *Adv Metab Disord* 1978; 9:29–48.

Wolff P: Vasomotor sensitivity to alcohol in diverse Mongoloid populations. *Am J Hum Genet* 1973; 25:193–199.

Wynder E: Dietary habits and cancer epidemiology. *Cancer* 1979; 43:1955–1961.

Yamase H, McNamara J: Geographic differences in the incidence of gallbladder disease: Influence of environment and ethnic background. *Am J Surg* 1972; 123:667–670.

chapter 5

Environmentally related variation

Climate, altitude, and diet are three major environmental stressors that affect large numbers of people. These factors produce bodily responses in one or more of three possible modes: phenotypic plasticity, irreversible response, and genetic response (Baker, Weiner, 1966).

Phenotypic plasticity refers to the reversible mechanisms that occur throughout the life of an individual. *Irreversible responses* are those that are environmentally labile only during the period of growth and development. *Genetic responses* are adaptations to an environmental stress that do not require exposure to the particular stress or that produce changes unique for the particular population when exposure occurs. Examples of these three modes of response follow.

Phenotypic plasticity is seen in the process of acclimatization to a different climate or altitude. After a few days or weeks of exposure to these stressors, individuals are able to withstand them better. The severity of the stress affects the individual's ability to respond. Most people can adjust to living at 1,525 m (5,000 ft) above sea level in a few weeks' time. The adjustment to 3,050 m (10,000 ft) is never as good. Similarly, people adjust to increased heat or cold, but extremes of temperature do not produce complete adjustment. Adjustment of dietary and fluid intake, improvement of physical fitness, change in work pace, and change in clothing all aid adjustment to a particular environmental stress. More calories are consumed by those living in constantly cool to cold environments than those living in warmer climates (Burton et al, 1940; Johnson, Kark, 1947a; Swain et al, 1949; Gounelle, 1949; Edholm, 1960). Voluntary intake increases to somewhere around 4,000 to 5,000 calories a day. There is a straight-line correlation between voluntary caloric intake and decrease in temperature: the colder the climate, the higher the caloric intake (Johnson, Kark, 1947b; Swain et al, 1949). Tolerance to cold is increased by frequent feedings (Mitchell et al, 1946; Kreiden, Buskirk, 1957). Subjects given supplemental feedings of either 600 or 1,200 calories before sleep had reduced periods of wakefulness, less subjective discomfort associated with cold, a significant increase in rectal and toe temperatures, and a significant increase in oxygen consumption, compared with control subjects (Mitchell et al, 1946; Kreiden, Buskirk, 1957).

Increased fluid intake is required when sweat production increases to adjust body temperature to heat stress. Physical fitness enables people to withstand these stresses better (Adams, Heberling, 1958; Budd, 1965; Scholander et al, 1958a; Andersen, Bolstad, Irving, 1960).

A particular environmental stress experienced during childhood produces irreversible changes that distinguish individuals in adulthood. This has been amply demonstrated in studies of cold injury to soldiers during war maneuvers. Men raised in warm climates suffered more cold injuries than those brought up in cold climates (Miller, Bjornson, 1962; Cook, 1956; Orr, Fainer, 1952; Newman, 1968).

Genetic responses can be distinguished from the other responses only with difficulty. Responses that do not involve phenotypic plasticity and are not due to childhood adaptation are considered genetic. Responses that appear unique to a particular group are also placed in this category.

Environmental influences on human biology have been intensively studied by three disciplines: physical anthropology, epidemiology, and physiology. Anthropologists, looking for populational adaptations to various environmental stresses as examples of the evolutionary process, have studied the distribution of specific adaptive characteristics in different races living in various climate and altitude zones. They have also been interested in the many diets that have evolved. Epidemiologists have looked for and found differences in people's abilities to withstand environmental stress. Physiologists have studied the musculoskeletal, integumentary, and cardiovascular systems to learn the many ways people react to outside influences. This chapter cites findings from the three disciplines to describe biologic responses to the environment—specifically, to climate extremes, high altitude, and diet.

CLIMATE

Three climatic factors affect health: cold, heat, and humidity. Air quality is also a pertinent climatic factor, but it will not be considered further here, because there are no data available yet on biologic variations in response to it.

Humans can adapt to climatic stress on a short- or long-term basis (Frisancho, 1981). For heat stress, short-term responses include excessive sweating, leading to sodium loss; excessive vasodilation, leading to lowered body temperature; and thirst. These responses are a form of phenotypic plasticity and occur in all healthy individuals.

Long-term adaptive responses occur after repeated exposure to a climatic stress over weeks, months, or years. Again, using the example of heat stress, after long exposure sweating decreases, so that sodium loss is decreased; vasodilation occurs but at a reduced rate, sparing the body temperature; and thirst responds to these adjustments. These long-term responses can be due to phenotypic plasticity or to irreversible responses developed previously.

Cold

Adaptation to cold increases an individual's ability to withstand cold stress safely and comfortably. Humans achieve this by several means: vasoconstriction alternating with vasodilation; shivering; and, in some instances, an increase in metabolic rate. Race, the degree of physical fitness, age, and body size and shape also affect cold adaptation.

The body's first response to cold stress is to conserve heat by constricting the peripheral blood vessels. This action preserves the core temperature but, if continued too long, can damage surface tissues by freezing them. Consequently, periodic vasodilation occurs. This alternating vasoconstriction-vasodilation sequence is called the *hunting reaction*. Cold-acclimatized individuals develop a very efficient hunting reaction that protects surface tissues from cold injury but prevents excessive heat loss from the body's core. Figure 5.1 shows a cold-acclimatized person's (Eskimo) response compared with that of nonacclimatized individuals (Black and White).

The hunting reaction is most effective in individuals raised in cold climates, those who have experienced cold stress on a regular basis later in life, and members of certain racial groups. The data on racial differences are confounded by geographic origin, but generally Eskimos, northern Indians, and Laplanders have the most effective hunting reactions (Meehan, 1955; Krog, Alvik, Lund-Larsen, 1969; Steegmann, 1977). Blacks generally have the least efficient responses; Whites have responses of intermediate effectiveness (Rennie, Adams, 1957).

Shivering is the second line of defense against cold stress. It works by increasing heat production. Shivering is not as efficient as the hunting reaction, because it *uses* energy instead of conserving it. People start shivering when the skin temperature cools to 28–30 C (82–86 F). There appears to be a racial difference in the shivering threshold. Whites and Eskimos start shivering when their skin temperatures lower to around 30 C (86 F), while Blacks do not shiver until their skin temperatures are down around 28 C (82 F) (Adams, Covino, 1958; Rennie, 1963). This racial difference is confounded by the effects of acclimatization. When any group of individuals is exposed to cold stress over several weeks or more, there is a decrease in shivering because of a rise in skin temperature. The higher skin temperature results from the more efficient use of the hunting reaction (Adams, Covino, 1958; Rennie, 1963).

Shivering elevates the metabolic rate temporarily, but there is some evidence that people who live under fairly constant cold stress have higher metabolic rates than those not so stressed. Eskimos in their native habitat had higher metabolic rates than White subjects with whom they were compared (Adams, Covino, 1958; Rennie, 1963; Milan, Evonuk, 1967). But, when white subjects were exposed to severe cold stress over several months, they also responded with increased metabolic rates (Scholander et al, 1958b). This increase occurs in both children and adults (Hill, Campbell, 1922). But it seems that the cold-stressed White subjects never achieve metabolic rates as high as those of Eskimos. This suggests a unique genetic adaptation in the Eskimos.

FIGURE 5.1 Finger temperature response during a cold test

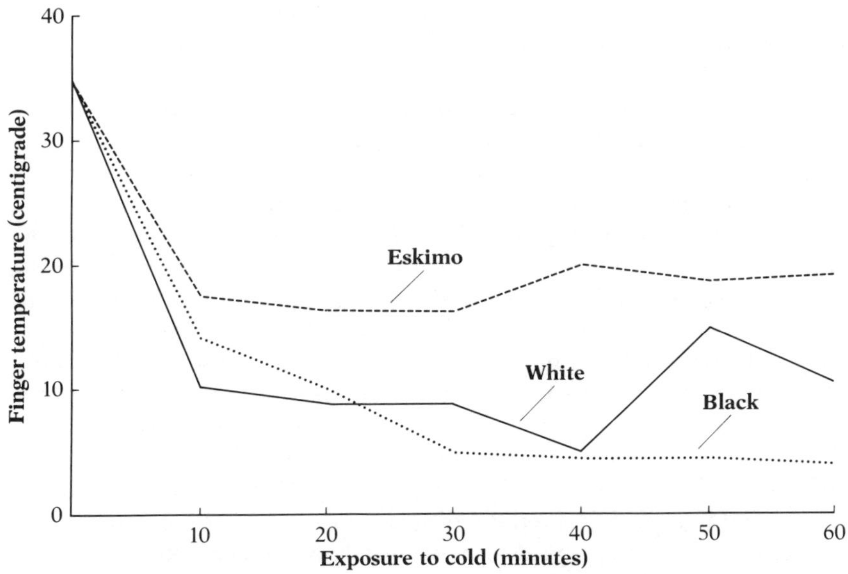

Sources: Compiled from Meehan, 1955; Rennie, Adams, 1957.

A few other factors affect human response to cold stress, but their effect is minor compared to the above responses. Physically fit individuals withstand cold stress better than those not in good shape. Fatter individuals, having more body insulation, conserve core temperature better than thinner individuals, but they experience similar subjective feelings of cold because their skin temperature falls as much as that of thinner people. Individuals at either extreme of the age spectrum are also more apt to suffer the effects of cold stress (Fox et al, 1973a; Frisancho, 1981).

The different responses to cold—hunting reaction, shivering, rise in metabolic rate—and the duration of cold stress are reflected in the differential susceptibility to cold injury observed in certain groups. A larger proportion of Blacks than Whites are prone to cold injury under the same cold exposure conditions (Orr, Fainer, 1952; Miller, Bjornson, 1962). Individuals raised in warm climates are also more susceptible to cold injury than those raised in cool to cold climates (Orr, Fainer, 1952; Miller, Bjornson, 1962). Whether race or lack of previous cold exposure is the greater risk factor in cold injury is subject to debate. But evidence of differences in the efficiency of the hunting reaction and the skin temperature at which shivering begins supports the notion that race is an underlying factor in any case.

The majority of cold injuries occur in wartime during sustained battles in the cold season of the year (Hanson, Goldman, 1969). Other injuries occur in hikers and climbers. Accidents and alcoholism also predispose to cold injury (Hanson, Goldman, 1969; Storrs, 1957).

Frostbite and hypothermia are the two severe consequences of cold exposure. Frostbite occurs when the skin temperature reaches the freezing point (0 C, 32 F). Hypothermia occurs when there is a lowering of core temperature and the skin temperature remains above freezing (Edholm, 1978). Symptoms develop when the core temperature decreases to 35 C (95 F). Consciousness is lost at around 30–32 C (87–90 F).

Nurses and physicians are more likely to see patients with hypothermia than with frostbite. A diagnosis of hypothermia or incipient hypothermia should be considered when dealing with the very young or very old during cold seasons, especially where housing is inadequate. Elderly individuals living alone are prone to hypothermia. Studies have shown that 17–24% of old people living at home have mouth temperatures at/or below 35.5 C (95.9 F) (Fox et al, 1973b). This is only 0.5 C from the threshold of hypothermia. The coldest elderly people were found to be living in the coldest dwellings, and, contrary to popular opinion, they did not feel much more uncomfortable than old people with higher temperatures. None of the cold subjects felt "too cold"; some felt "on the cool side" (Fox et al, 1973b, p 23).

Health care providers need to be especially concerned about the old individuals they see "who tend to have a low mouth temperature, are living in poorly heated accommodations, and yet say they don't feel the cold. These are a group of old people at risk of developing hypothermia, if for any reason they become immobilized in cold conditions" (Fox et al, 1973b, p. 24).

Heat

Within a temperature range of 20–24 C (68–75 F), approximately two thirds of the excess body heat is lost through the process of radiation (Frisancho, 1981). As the temperature rises, less heat is lost this way, and evaporation of sweat becomes the major mechanism for heat loss. At temperatures over 35 C (95 F)—or, if an individual is working hard, at even lower air temperatures—up to 90% of heat loss is through the evaporation of sweat. If the humidity is high when the air temperature is high, the body finds it difficult to maintain its normal temperature because of the lessened efficiency of sweat evaporation. As more sweat is produced than can be evaporated, the skin becomes wet. This overproduction of sweat at high air temperatures wastes body fluid. Under prolonged heat stress, fluid and mineral replacement is necessary, or dehydration and a seriously raised body temperature will result.

The body adjusts to heat stress fairly rapidly. Within four days the body temperature stabilizes, and within two weeks most people adjust their sweating rates (D Mitchell et al, 1976). But in the first few days of exposure to high heat, especially when physical work is being done, body temperature can rise dramatically. Rectal temperatures of 38 C (100.4 F) are common; under hard work, healthy men may increase their rectal temperatures to 40 C (104 F) (Senay, Mitchell, Wyndham, 1976). Symptoms of hyperthermia—syncope, decrease in sweating, irregular pulse, severe malaise—will generally cause the individual to stop exertion before damage is done.

Upon first exposure to heat stress, the body responds with a rise in temperature and heart rate that indicates circulatory stress. If the heat stress continues, the sweat rate increases over the next few days. Simultaneously the effectiveness of vasodilation of the superficial blood vessels increases. These two phenomena are accompanied by an increase in plasma volume from a transfer of interstitial protein and water into the vascular system (Senay, Mitchell, Wyndham, 1976). As these body responses come into play, the circulatory stress is relieved, allowing the heart rate to go down. The body temperature also decreases as the cooling mechanisms respond. If heat stress continues, the body adjusts the sweat rate to a lower, more conservative level; within two to three weeks, the body achieves equilibrium for the higher temperature. At equilibrium, the person again feels comfortable.

Sweat production is an obvious manifestation of the body's response to heat. People can influence the effects of sweating by consciously replacing fluids, electrolytes, and minerals. Under conditions of heat stress, they must make a deliberate effort to replace the lost sweat constituents, because the sensation of thirst is not a good indication of body needs. When sweating is profuse, the body responds with thirst only until the body fluids have reached the point of isotonic dilution (Allan, Wilson, 1971; Consolazio et al, 1963). This state can be reached before the body is fully hydrated. When this occurs, the dehydrated state persists until food is consumed that replaces the lost electrolytes and minerals. At this point the thirst sensation is activated. Until the body is fully hydrated, the body temperature remains elevated.

The indiscriminate emphasis on low-sodium diets may increase the possibility of dehydration in individuals who work hard and consequently sweat heavily. Such individuals should avoid low-sodium diets and probably should add an electrolyte-mineral replacement to their fluid intake. Several replacement packets and solutions are now available in food markets and sporting goods stores.

Susceptibility to heat stress is influenced by age, race, body build, body fat, climatic experience, and possibly sex. Persons at either extreme of the age spectrum suffer from the heat more than other people. Both groups are more prone to dehydration; body temperature regulation and sweating efficiency are less than perfectly functional at these ages (Wagner et al, 1972). Infants cannot adjust their clothing, move to a cooler location, or take a drink when they are thirsty. It takes several years before the sweating mechanism becomes fully functional. Even preadolescent children are not as heat tolerant as adolescents and young adults, and after 50 years of age a decline in heat tolerance is noticeable (Wagner et al, 1972).

Certain races appear to withstand heat stress better than others, but the racial effect is confounded by climatic exposure. Persons brought up in hot climates are better able to tolerate heat than those raised in cooler regions (Frisancho, 1981). Also, individuals who later reside in hot climates acclimatize and consequently withstand heat better than newcomers. Even though studies have been designed that attempt to control for climatic origin and experience, the existence of true racial differences is still not certain. But, in general, Blacks, some southwestern American Indians, Asian Indians, and Australian Aborigines withstand heat better than Whites (Riggs, Sargent, 1964; Frisancho, 1981). They have lower sweat rates than Whites with similar body temperatures, indicating more efficient sweat production (Dill et al, 1983).

Part of the climatic influence on heat tolerance is related to the effect of heat on the development of body build during childhood. Children who grow up in hot climates are more slender than their counterparts raised in more temperate areas (Eveleth, 1966). Such children are the same height, but weigh less, having thinner arms and legs. The question arises whether these differences, like the racial differences in body proportion and rates of growth, are due to race or to climate.

People with slender extremities have more surface area per body weight; this gives them a proportionately larger surface for sweat production and evaporation compared with individuals who are heavily muscled or obese. This explains why heavier people suffer more from the heat. Their body temperatures are not as easily adjusted because their sweating mechanism is less efficient (Bar-Or, Lundegren, Buskirk, 1969; Shvartz, Saar, Benor, 1973).

The sex differences in amount of body fat may explain some of the sex difference in heat tolerance, but other studies show virtually identical abilities of males and females to withstand heat (Bar-Or, Lundegren, Buskirk, 1969; Dill, Yousef, Nelson, 1973; Kamon, Avellini, 1976). Several of the studies where sex differences were noted use a combination of physical work and heat stress; perhaps the women were not as capable of sustaining the physical work component as were the men with whom they were compared (Dill, Yousef, Nelson, 1973; Frisancho, 1981).

The age factor is reflected in heat casualties recorded during the summer heat wave of 1980. There were an estimated 1,265 deaths due to the heat, and the heat stroke rate in persons 65 years of age and older was 13 times higher than in other age groups (Donnell et al, 1981). Heat stroke, one of the three illnesses associated with heat stress, usually occurs during extended heat waves when the temperature and humidity are high. The victim sweats continuously for several days until the sweat mechanism fails due to insufficient fluid and electrolyte replacement. Failure of the sweat mechanism leads to a dangerously high body temperature (Knochel, 1974). Older people are at risk for heat stroke.

Another type of heat illness is heat cramp, severe muscle cramps. This condition is seen most often in acclimatized individuals doing heavy physical work who replace fluids but not electrolytes (Knochel, 1974). Construction workers and athletes are likely victims of this malady.

The most common heat illness is heat exhaustion (Knochel, 1974). It occurs in two subtypes: water depletion and salt depletion. In the former, individuals become dehydrated from not drinking enough water; in the latter, water is replaced but electrolytes are not. The main differences between salt-depletion heat exhaustion and heat cramps are these: heat exhaustion occurs in nonacclimatized individuals and the syndrome has systemic symptoms—thirst, malaise, discomfort, anxiety, and impaired judgment; heat cramps occur in acclimatized individuals. Both subtypes of heat exhaustion have the same set of symptoms (Knochel, 1974).

During hot spells, health workers need to be alert to incipient heat illnesses. They should counsel those caring for infants and the elderly on the increased susceptibility to heat of these two age groups. General health education should include information about adequate replacement of water and electrolytes and other measures to avoid heat illness—taking it easy until acclimatized, being careful to shield youngsters and oldsters from the full effects of heat, and being aware of who is at greatest risk for heat illness.

ALTITUDE

Many of the biologic correlates of high altitude residency were formerly considered the result of genetics. Most of these differences have now been properly assigned to the effect of high altitude. Living at high altitude influences growth patterns, lung function, blood values, and blood pressure.

Most high altitude research has been done on populations living at the upper limits of human habitation, such as in Leadville, Colorado, at 3,050 m (10,000 ft) and in villages in the Peruvian Andes 3,650–4,300 m (12,000–14,000 ft). Most health care workers and the people they serve come from much lower altitudes, but in many large population centers in the Rocky Mountain states milder effects of altitude are evident. Denver, Colorado, at 1,525 m (5,000 ft) and Santa Fe, New Mexico, at 2,135 m (7,000 ft) are two examples of cities at moderately high altitudes.

As altitude increases, the amount of oxygen inhaled at each breath decreases. Oxygen is 21% of the atmosphere at all altitudes, but the gas molecules are compressed at sea level and at low altitudes in general. As altitude increases, the gas molecules become less concentrated; consequently less oxygen is available in each breath taken. This is the cause of high-altitude stress—hypoxia.

The differences in oxygen concentration affect people as they leave sea level and ascend to higher elevations. Individuals notice short-term effects, such as shortness of breath, headache, and accelerated pulse, when they travel to higher altitudes. These short-term effects occur in some people at altitudes as low as 1,525 m (5,000 ft). They usually disappear within eight days (Frisancho, 1981).

Long-term effects are not noticeable until moderate altitudes are reached—1,525 m (5,000 ft)—and the effects are subtle until people are at relatively high altitudes—3,050 m (10,000 ft). For instance, birth weights decrease and placental weights increase at higher altitudes. Children grow more slowly at high altitudes, but their growth period is longer, to compensate. See chapter 3 for more details.

Lung capacity and chest size are greater in persons living at high altitudes. This occurs whatever the race in individuals born and raised at high altitudes (Mueller et al, 1978; Beall, 1982). Individuals who migrate to high-altitude locations experience an increase in lung capacity and chest size, but less increase than that shown by individuals native to the area (Mueller et al, 1979). The laboratory measurements that show these differences are the ventilatory values: peak flow rate (PFR), forced expiratory volume at 1 and 3 seconds (FEV_1 and FEV_3), and forced vital capacity (FVC).

Hypoxic stress also produces an adjustment in the oxygen-carrying capacity of the blood. Hematocrits, hemoglobin levels, and red blood cell (RBC) counts increase as altitude increases (Frisancho, 1981). The blood volume increases minimally; consequently the blood is thicker. This polycythemia puts some strain on the heart, which helps explain the higher pulse rate that occurs during the first weeks of acclimatization.

The rate of increase in hemoglobin level, hematocrit, and RBC count is gradual at altitudes up to about 2,700 m (9,000 ft); then the rate rises steeply (Arnaud, Quilici, Riviere, 1981). Measurably higher values occur around 1,525 m (5,000 ft) but the increases are dramatic by 3,350 m (11,000 ft). At the higher elevations, the RBC count is 5–8 million compared with sea level values of 4.5–6 million; the hemoglobin level is 16–20 g versus sea level values of 14–18 g; and the hematocrit is 48–60% versus 40–54% at sea level (Garruto, 1976; Arnaud, Quilici, Riviere, 1981; Arnaud et al, 1979).

There is some controversy about whether blood pressure is lower and hypertensive heart disease decreases at high altitudes. Some studies say yes (Hurtado, 1960; Penalza et al, 1963; Marticorena et al, 1969; Mortimer, Monson, MacMahon, 1977); others say no (Morton, Davids, Lichty, 1964; Moore et al, 1982). Three of the four studies that indicate a decrease in blood pressure and hypertension at high altitudes were done in Peru on healthy natives or healthy White male workers. Both studies that showed no decrease were done in Colorado on mostly White populations in which sick individuals were included. The Colorado population lives a considerably different life-style from the Peruvian population. Because of the socioeconomic factors involved in blood pressure and hypertension, studies that control for such factors are needed before the controversy can be settled.

DIET

Dietary differences that relate to health arise mainly in the amount and types of food eaten and in specific eating customs. These factors help produce some of the biologic variations seen in populations. Dietary patterns will be viewed here in terms of how they promote or inhibit health.

Among people who have recently migrated to the United States, or who still retain an identity with their country of origin, there often is a diet dependent on a particular staple. For instance, Oriental diets are based on rice or millet. Many African diets use millet. Many American Indian and Chicano diets are based on corn and beans. A few tropic groups base their diet on manioc (cassava). Irish use the potato. Scots depend on oatmeal. Riverine Eskimos in Alaska depend on fish.

Dietary counseling in the United States has generally been based on the "four basic" food groups. These food groups are appropriate for Americans eating a standard American diet but are not necessarily applicable to individuals eating different cuisines. For instance, the milk group is not applicable to many cuisines. To avoid ethnocentrism in dietary counseling, it is better to use the broader categories of proteins, fats, carbohydrates, vitamins, minerals, water, and perhaps roughage. The body needs certain amounts of each of these categories, and calculation of an individual's needs can be derived from the United States Recommended Dietary Allowances (US RDA) or the recommendations of the World Health Organization. These RDAs are useful guidelines but are designed to cover the needs of 95% of the population; they are set at approximately two standard deviations above the mean requirement (Hegsted, 1972). Thus, many individuals who do not reach the recommended levels may nevertheless be adequately nourished.

Since the RDAs may not always help when diets are being assessed, other criteria need to be used. For instance, are children growing at rates comparable to a healthy cohort? Are the children or adults active and alert? Do they seem to stay healthy or do they constantly get sick? Dietary determinations can be made on the basis of physical assessment once it is determined that individuals get some of each dietary component, although not necessarily the high levels of the RDAs.

Many of the cuisines of the world can be distinguished by which food substance is the dietary staple. This is an easier way of looking at diet patterns than trying to learn the diets of the various groups being served by a health agency. Each of the cuisines, focusing on its own dietary staple, requires a different approach for nutrition counseling. For instance, counseling individuals using an American dietary pattern—which is wheat-based but also rather diversified—is quite different from counseling individuals on rice-based diets.

In developed nations, although there is less dependence on dietary staples, there is usually an underlying staple, such as wheat in the standard American diet. Generally, the poorer the family, the more it depends on the dietary staple.

Information on specific dietary staples, and what health care providers should be alert for in cuisines using them, follows. Also, some specific eating habits and their nutritional implications are addressed. Finally, some consequences of overnutrition and undernutrition are examined.

Dietary Staples

Corn (maize), millet, rice, and wheat are the dietary staples used by a large proportion of the world's populations. Fish and sea mammals, manioc (cassava), and white or sweet potatoes are less widely used staples. These less-used dietary staples will not be considered further here; suffice it to say that dietary counselors should adjust the advice they give to people using such foods.

Corn The main concern with corn-based diets is that corn lacks lysine and tryptophan, two essential amino acids (Haas, Harrison, 1977). Corn is also low in niacin, and most of what is present occurs in an unusable form (Haas, Harrison, 1977). Thus corn-based diets have a potential to produce protein malnutrition and pellagra. Both conditions are rare in the Western Hemisphere because of the traditional use of beans in corn-based diets. Because beans and corn have complementary amino acid compositions, protein malnutrition does not occur when they are used together. The niacin deficiency is helped by the common practice of treating corn with some sort of alkali—lye, lime, or wood ashes. This changes the niacin content into a usable form. If corn and bean diets contain a variety of fruits and vegetables, as they often do, they are adequate when sufficient calories are ingested. Meat, milk, and wheat products are not necessary for adequate nutrition under this dietary regimen.

Rice Rice-based diets can be low in protein, fats, and water-soluble vitamins (Dwyer, Oiso, 1981). It is estimated that the traditional Japanese diet is 83% carbohydrate, 11% protein, and 6% fat (Dwyer, Oiso, 1981). The quantity of protein is generally more of a problem in the Oriental diet than its quality. The traditional practice of serving over rice such items as soybean products, dried or fresh fish, fish sauce, and vegetables adds protein but sometimes not enough in poor families. When parboiled rice is substituted for milled and washed rice, water-soluble vitamins are saved. Unfortunately most Orientals do not like parboiled rice, although it is used in India. Economically deprived groups on rice-based diets need to be monitored for protein malnutrition and vitamin deficiency. Fats should also be monitored, as they are generally low also.

Millet Millet-based diets have the same problems as rice-based diets, plus a unique problem—possible contribution to the causation of goiters (Osman, 1981). This probably is not a problem in the Western Hemisphere, but millet has been implicated as a goitrogenic substance in the Sudan area of Africa and the millet-growing regions of India.

Wheat Wheat-based diets can be low in protein, vitamins, and minerals but are rarely low in fats or carbohydrates (Haas, Harrison, 1977). Dietary counseling of people on wheat-based diets can include information on the four basic food groups, except for individuals or groups who are deficient in lactase. For such people, the milk groups should be modified to include only aged cheese and soured milk products, such as yogurt. Most if not all of the protein and calcium the milk group is supposed to supply can be obtained by using foods from the other three groups. The basic four food groups were designed for wheat-based diets that include milk, not for any other diet.

Eating Habits

Several food habits also influence nutrition and health. Three of the more frequently encountered ones (pica, feeding frequency, and milk ingestion) are described and their nutritional implications explained.

Pica Pica is a chronic, compulsive desire to eat nonfoods (clay, starch), unusual combinations of foods (pickles and ice cream), or ordinary foods in unusually large amounts or at odd times of the day or night (Hunter, 1973; Chandra, Rosner, 1973; Perry, 1977; Lackey, 1978). The cause of pica is still debated, but many investigators are convinced that it meets some biologic need rather than a strictly cultural or psychologic need (Lackey, 1978). Iron-deficiency anemia and the need for more calories in the diet are associated with the practice (Lackey, 1978; Hunter, 1973). Frequently, specific nutrient supplementation will halt the craving. The food cravings of pregnancy are also often relieved by iron supplementation.

Contrary to common opinion, pica is not racially specific. One survey of Black and White women in Tennessee indicated that pica occurred irrespective of race, age, or educational attainment (Lackey, 1978).

Feeding frequency The number of times food is eaten during a day has some effect on utilization of specific nutrients. There is some evidence that people who eat more than three meals per day are less fat and more muscular than those who confine themselves to breakfast, lunch, and dinner (Cohn, Joseph, Allweiss, 1962). This difference in feeding frequency presupposes that a similar numbers of calories are ingested in the two patterns. In the United States, Canada, and England, for instance, the practice of eating more than three times a day is fairly common (Cohn, Joseph, Allweis, 1962; Johnston, Williams, Weldon, 1977; Thomas, 1982). The morning coffee break is well established in homes and business firms. Children frequently snack after school, and evening snacks are also popular. Rather than consider these practices poor nutrition, one group of researchers suggests that health workers encourage them but suggest the substitution or addition of nutritious foods—high-protein foods, fruits, and vegetables (Cohn, Joseph, Allweiss, 1962). Parents' common concern that snacks will spoil a child's appetite may be well founded. Perhaps snacking keeps the appetite less than voracious, a good thing in today's somewhat overweight American society.

Milk ingestion Drinking milk and eating milk products after weaning are common practices of many in the White race. Consumption of soured milk products and cheeses is also practiced by many Whites and occasionally by members of other races. While many groups consider this practice somewhat abnormal and even nauseating, ingesting milk products can be beneficial. Milk contains a protein with well-balanced amino acids. It also contributes some fat and carbohydrate in the form of lactose. Many Whites have inherited the ability to continue digesting milk sugar after weaning. This inherited mutation keeps the lactose enzyme active after the weaning period when it normally disappears. The mutation and its consequences—the ability to drink the milk of cows and goats—is an advantage in some ways. It allows a person to drink a fairly inexpensive, well-balanced protein, thus satisfying thirst and part of the protein and carbohydrate needs at the same time.

Consequences of Overnutrition and Undernutrition

The problems of over- and undernutrition are well known. Cardiovascular diseases, diabetes, and obesity are examples of disorders related to overnutrition. Severe growth retardation (kwashiorkor and marasmus), less severe chronic growth retardation, and specific deficiency diseases (beriberi, pellagra, and scurvy) are examples of disorders related to undernutrition. Severe undernutrition problems are rarely seen in industrialized countries. Chronic growth retardation is more common, occurring in poor and chronically ill children. Growth retardation is discussed in chapter 3. Severe undernutrition is not mentioned further.

The overnutrition diseases mimic biologic variation; they occur in certain racial groups more than in others. Three such diseases—hypertension, diabetes, and obesity—are predictably present in certain racial groups. These and other overnutrition-related diseases are more often due to life-style differences and recent life-style changes than to innate biologic differences between groups.

Hypertension Hypertension is considered a disease of aging. It occurs most often in urban areas and industrial countries. In the United States, Blacks have more hypertension than Whites. The highest rates of hypertension occur in Japan and Taiwan, two highly urban, industrial countries. It is also more common in people who have recently migrated or changed from less to more urban and industrial conditions. Obesity and salt ingestion are implicated in the causation of hypertension (Prior, 1971).

The common thread running through the variables of urbanization, industrialization, and migration is stress, which is known to increase blood pressure. In addition to stress, the other variables generally produce a change from physically active to less active work, a ready availability of foodstuffs, and consequently a tendency to put on weight. Obesity is also associated with hypertension. Many individuals who develop hypertension under these circumstances would probably never manifest it in a less-stressed milieu, with less overeating and more physical activity.

The distribution of hypertension leads me to suspect that, as groups become familiar with the stresses of urban and industrial life, and as they gain socioeconomic status, they make life-style adjustments that reduce the rate of hypertension. This is borne out by data on hypertension rates mentioned in the section on aging in chapter 3. Blacks who had positions of greater status and monetary reward had less hypertension than their less fortunate cohorts (Gordon, Devine, 1973).

Because of the greater ability of Blacks to moderate their sweat output under the stimulation of heat stress, they lose less salt during sweating than do Whites. Perhaps Black individuals need to be more careful than Whites about regulating salt ingestion—using enough to prevent heat problems, but not so much that they cause a rise in blood pressure.

Diabetes and obesity Diabetes and obesity are closely entwined health problems. Groups with a high frequency of both are identified in chapter 4. The common denominator in these groups seems to be a recent change in diet and life-style. Without considering the many factors that undoubtedly affect the production of overt symptoms of diabetes, it is appropriate to counsel obese individuals from groups experiencing rapid dietary and life-style changes to reduce their excess weight by an increase in exercise and a modification of diet. This is easy to say, but unfortunately results are hard to achieve. Chapter 8 on the Sioux Indians includes some suggestions on appropriate preventive techniques from a practitioner who knows the specific population.

Table 5.1 compares data on diabetes, hypertension, and obesity in several Pacific Island groups. It illustrates the multiple factors involved in the production of disease as life-styles and diets change.

The reasons diabetes is so common in certain groups is not known; one hypothesis is that a diabetic tendency may be advantageous for people who experience famine cycles (Neel, 1963). See the discussion in chapter 4.

TABLE 5.1 Comparison of diets and diseases in four Pacific populations

Diet factor or disease	Population group			
	Pukapuka	Rarotonga	New Zealand Maori	New Zealand "Europeans"
Calories (daily intake)	1,800 (ages 40–69)	2,100 (ages 40–69)	2,560 (all ages)	2,270 (all ages)
Fat (daily intake)	70 g	63 g	125 g	99 g
Sucrose (daily intake)	9 g	35 g	71 g	61 g
Salt (daily intake)	50–70 mEq	120–140 mEq	188 mEq	160 mEq
Cholesterol (mean serum cholesterol)				
Males (ages 50–59)	187 mg/ 100 mL	212 mg/ 100 mL	232 mg/ 100 mL	275 mg/ 100 mL
Females (ages 50–59)	200 mg/ 100 mL	235 mg/ 100 mL	225 mg/ 100 mL	268 mg/ 100 mL
Obese (120–139% of recommended weight)				
Males (ages 30 and over)	7%	8%	18.6%	6.7%
Females (ages 30 and over)	21%	22%	21.8%	17%
Grossly obese (140% of recommended weight)				
Males (ages 30 and over)	0.7%	3.7%	9.1%	1.2%
Females (ages 30 and over)	2.2%	25%	21.4%	3.7%
Mean weights at age 40				
Males	152 lbs	180 lbs	NA	NA
Females	148 lbs	189 lbs	NA	NA
Gout (incidence in ages 30 and over)				
Males	7%	3.7%	13.3%	2.4%
Females	0.0%	0.0%	2.5%	0.0%

Continued on next page.

TABLE 5.1, continued

Diet factor or disease	Population group			
	Pukapuka	Rarotonga	New Zealand Maori	New Zealand "Europeans"
Diabetes (incidence in ages 30 and over)				
Males	2%	6.2%	11.9%	1.8%
Females	4.4%	9.5%	9.3%	4.8%
High blood pressure, above 160/95 (incidence in ages 30 and over)				
Males	2%	21%	16.9%	22%
Females	4.4%	36%	25%	26%
Ischemic heart disease (incidence in ages 30 and over)				
Males	2%	4.3%	4.9%	4.2%
Females	1.4%	5.5%	8.9%	9.6%

Source: Adapted from Prior, 1971, pp 8–9.

Summary

Much of the biologic variation in human beings that was previously considered racially specific is now known to be produced by the effects of heat, cold, altitude, and dietary experience. Body responses to environmental stress are generally protective; but health care providers need to recognize when preventive or restrictive measures should be taken to reduce these stresses.

The best approach to dietary counseling is from the perspective of the major dietary staple, rather than the basic four food groups, so that the advice health providers give is compatible with the diets of their clients.

REFERENCES

Adams T, Covino B: Racial variations to a standardized cold stress. *J App Physiol* 1958; 12:9–12.

Adams T, Heberling E: Human physiological responses to a standardized cold stress as modified by physical fitness. *J Appl Physiol* 1958; 13:226–230.

Allan J, Wilson C: Influence of acclimatization on sweat sodium concentration. *J Appl Physiol* 1971; 30(5):708–712.

Andersen K, Bolstad Y, Irving L: Physical fitness of Arctic Indians. *J Appl Physiol* 1960; 15:645–648.

Arnaud J, Quilici J, Riviere G: High-altitude haematology: Quechua-Aymara comparisons. *Ann Hum Biol* 1981; 8(6):573–578.

Arnaud J et al: Methaemoglobin and erythrocyte reducing systems in high-altitude natives. *Ann Hum Biol* 1979; 6(6):585–592.

Baker P, Weiner J (editors): *The Biology of Human Adaptability*. Clarendon Press, 1966.

Bar-Or O, Lundegren H, Buskirk E: Heat tolerance of exercising obese and lean women. *J Appl Physiol* 1969; 26(4):403–409.

Beall C: Comparison of chest morphology in high-altitude Asian and Andean populations. *Hum Biol* 1982; 54(1):145–163.

Budd G: Effects of cold exposure and exercise in a wet cold Antarctic climate. *J Appl Physiol* 1965; 20:417–422.

Burton A et al: Slow adaptations in the heat exchanges of man to changed climatic conditions. *Am J Physiol* 1940; 129:84–101.

Chandra P, Rosner F: Olives-craving in iron deficiency anemia. *Ann Intern Med* 1973; 78(6):973–974.

Cohn C, Joseph D, Allweiss M: Nutritional effects of feeding frequency. *Am J Clin Nutr* 1962; 11:356–361.

Consolazio C et al: Excretion of sodium, potassium, magnesium, and iron in human sweat and the relation of each to balance and requirements. *J Nutr* 1963; 79:407–415.

Cook E: *The Epidemiology of Frostbite*. Naval Medical Research Institute, 1956.

Dill D, Yousef M, Nelson J: Responses of men and women to two-hour walks in desert heat. *J Appl Physiol* 1973; 35(2):231–235.

Dill D et al: Volume and composition of hand sweat of White and Black men and women in desert walks. *Am J Phys Anthropol* 1983; 61(1):67–73.

Donnell H et al: Heatstroke—United States, 1980. *Morbid Mortal Weekly Rep* 1981; 30(23):277–279.

Dwyer J, Oiso T: Changing food patterns. Pages 509–525 in: *Nutrition in Health and Disease and International Development*. Symposia from XIIth Int Cong Nutr. Liss, 1981.

Edholm O: *Man—Hot and Cold*. Camelot Press, 1978.

Edholm O: Polar physiology. *Fed Proc* 1960; 19:39.

Eveleth P: The effects of climate on growth. *Ann NY Acad Sci* 1966; 134(2):750–759.

Fox R et al: Body temperatures in the elderly: A national study of physiological, social, and environmental conditions. *Br Med J* (January) 1973a; 27:200–206.

Fox R et al: Problem of the old and the cold. *Br Med J* (January) 1973b; 6:21–24.

Frisancho A: *Human Adaptation*. University of Michigan Press, 1981.

Garruto R: Hematology. Pages 261–282 in: *Man in the Andes*. Baker P, Little M (editors). Dowden, Hutchinson, & Ross, 1976.

Gordon T, Devine B: Hypertension and hypertensive heart disease in adults. *Vital Health Stat* 1973; 11(13):1–62.

Gounelle H: Nutrition and the very cold climate. *Ann Nutr* 1949; 3:471–489.

Gupte O, Joshi M, Dave S: Prevalence of diabetes in India. *Adv Metab Disord* 1978; 9:147–166.

Haas J, Harrison G: Nutritional anthropology and biological adaptation. *Annu Rev Anthropol* 1977; 6:69–101.

Hammon R, Bennett P, Miller M: Incidence of diabetes among the Pima Indians. *Adv Metab Disord* 1978; 9:49–63.

Hanson H, Goldman R: Cold injury in man. *Milit Med* 1969; 134:1307–1316.

Hegsted D: Problems in the use and interpretation of the recommended dietary allowances. *Ecol Food Nutr* 1972; 1:255–265.

Hill L, Campbell J: Metabolism of children undergoing open air treatment, heliotherapy, and balneotherapy. *Br Med J* 1922; 1:301–303.

Hunter J: Geophagy in Africa and the United States. *Geograph Rev* 1973; 63:171–195.

Hurtado A: Some clinical aspects of life at high altitudes. *Ann Intern Med* 1960; 53(2):247–258.

Johnson R, Kark R: Environment and caloric requirements. *Fed Proc* 1947a; 6:138.

Johnson R, Kark R: Environment and food intake in man. *Science* 1947b; 105:378–379.

Johnston J, Williams C, Weldon K: Nutrient intake and meal patterns of Micmac Indian and Caucasian women in Shubenacadie, NS. *Can Med Assoc J* 1977; 116:1356–1359.

Kamon E, Avellini B: Physiologic limits to work in the heat and evaporative coefficient for women. *J Appl Physiol* 1976; 41(1):71–76.

Kawate R et al: Preliminary studies of the prevalence and mortality of *diabetes mellitus* in Japanese in Japan and on the island of Hawaii. *Adv Metab Disord* 1978; 9:202–224.

Knochel J: Environmental heat illness. *Arch Intern Med* 1974; 133:841–864.

Kreiden M, Buskirk E: Supplemental feeding and thermal comfort during sleep in the cold. *J Appl Physiol* 1957; 11:339–343.

Krog J, Alvik N, Lund-Larsen K: Investigations of the circulatory effects of submersion of the hand in ice water in the Finnish Lapps, the "Skolts." *Fed Proc* 1969; 28:1135–1137.

Lackey C: Pica—a nutritional anthropology concern. Pages 121–129 in: *The Anthropology of Health*. Bauwens E (editor). Mosby, 1978.

Marticorena E et al: Systemic blood pressure in White men born at sea level: Changes after long residence at high altitudes. *Am J Cardiol* 1969; 23:364–368.

Meehan J: Individual and racial variations in a vascular response to a cold stimulus. *Milit Med* 1955; 116:330–334.

Milan F, Evonuk E: Oxygen consumption and body temperature of Eskimos during sleep. *J Appl Physiol* 1967; 22:565–567.

Miller D, Bjornson D: An investigation of cold injured soldiers in Alaska. *Milit Med* 1962; 127:247–252.

Mitchell D et al: Acclimatization in a hot, humid environment: Energy exchange, body temperature, and sweating. *J Appl Physiol* 1976; 40:768–778.

Mitchell H et al: The tolerance of man to cold as affected by dietary modification. *Am J Physiol* 1946; 146:84–96.

Moore L et al: Incidence of pregnancy-induced hypertension is increased among Colorado residents at high altitude. *Am J Obstet Gynecol* 1982; 144(4):423–429.

Mortimer E, Monson R, MacMahon B: Reduction in mortality from coronary heart disease in men residing at high altitude. *N Engl J Med* 1977; 296(11):581–585.

Morton W, Davids D, Lichty J: Mortality from heart disease at high altitude. *Arch Environ Health* 1946; 9:21–24.

Mueller W et al: A multinational Andean genetic and health program: VI. Physiological measurements of lung function in an hypoxic environment. *Hum Biol* 1978; 50(4):489–514.

Mueller W et al: A multinational Andean genetic and health program: VIII. Lung function changes with migration between altitudes. *Am J Phys Anthropol* 1979; 51(2):183–196.

Neel J: Diabetes mellitus: A "thrifty" genotype rendered detrimental by "progress"? *Am J Hum Genet* 1963; 14(4):353–363.

Newman R: Cold acclimation in Puerto Ricans. *J Appl Phsyiol* 1968; 25:277–282.

Orr K, Fainer D: Cold injuries in Korea during winter of 1950–51. *Medicine* 1952; 31:177–220.

Osman A: Bulrush millet (Pennisetum typhoides) a contributory factor to the endemicity of goitre in Western Sudan. *Ecol Food Nutri* 1981; 11:121–128.

Penalza D et al: Pulmonary hypertension in healthy men born and living at high altitudes. *Am J Cardiol* 1963; 11:150–157.

Perry M: Cautopyreiophagia. *N Engl J Med* 1977; 296:824.

Prior I: The price of civilization. *Nutr Today* (July/August) 1971; 6:2–11.

Prior I et al: The relationships of diabetes blood lipids, and uric acid levels in Polynesians. *Adv Metab Disord* 1978; 9:242–262.

Rennie D: Comparison of nonacclimatized Americans and Alaskan Eskimos. *Fed Proc* 1963; 22(3):828–830.

Rennie D, Adams T: Comparative thermoregulatory responses of Negroes and White persons to acute cold stress. *J Appl Physiol* 1957; 11:201–204.

Riggs S, Sargent F: Physiological regulation in moist heat by young American Negro and White males. *Hum Biol* 1964; 36(4):339–353.

Scholander P et al: Cold adaption in Australian Aborigines. *J Appl Physiol* 1958a; 13:211–213.

Scholander P et al: Metabolic acclimatization to cold in man. *J Appl Physiol* 1958b; 12:1–8.

Senay L, Mitchell D, Wyndham C: Acclimatization in a hot, humid environment: Body fluid adjustments. *J Appl Physiol* 1976; 40(5):786–796.

Shvartz E, Saar E, Benor D: Physique and heat tolerance in hot-dry and hot-humid environments. *J Appl Physiol* 1973; 34(6):799–803.

Steegmann A: Finger temperatures during work in natural cold: The northern Ojibwa. *Hum Biol* 1977; 49(3):349–362.

Storrs H: Frostbite. *North Med* 1957; 56:816–818.

Swain H et al: Food consumption of soldiers in a subarctic climate. *J Nutr* 1949; 38:63–72.

Thomas J: Food habits of the majority: Evolution of the current UK pattern. *Proc Nutr Soc* 1982; 41:211–228.

Wagner J et al: Heat tolerance and acclimatization to work in the heat in relation to age. *J Appl Physiol* 1972; 33(5):616–622.

West K: Diabetes in American Indians. *Adv Metab Disord* 1978; 9:29–48.

chapter 6

Nervous system variation

Research findings are scarce on the topic of biologic variation in the nervous system. A good deal has been written on the subject, but very little is scientific. The majority of accounts are speculations, often from biased viewpoints.

This chapter reviews the small amounts of available data on biologic variation by race in the five senses, in nervous system disease susceptibility, and in brain development and mental illness. Mental functioning by race has been studied extensively, but only those few studies where biologic variation can be demonstrated are reviewed. Nervous system functioning by sex and by age are dealt with in chapters 7 and 3 respectively.

THE FIVE SENSES

Data from the racial perspective are available for vision, hearing, and smell. The sense of taste has been narrowly studied as it relates to phenylthiocarbamide (PTC) tasting, but little else is available. The sense of touch has been poorly studied; much has been written on cultural differences in pain perception and response, but few data have been produced from a biologic perspective.

Vision

Acuity Visual acuity shows differences by race. Surveys done before the 1960s indicated an acuity difference between Blacks and Whites, favoring

Blacks. Studies done since that time have clarified the difference. What appears to be the case is that Blacks generally have better uncorrected vision than Whites in the 20/50 or better range (93.2% of Black men versus 83.7% of White men are in this range), but at 20/20 or better, the races are very close (59.9% for Black men and 57.3% for White men) (Roberts, 1964). The use of less precise tests a few decades ago probably led to the impression that Blacks had much better visual acuity than Whites.

A National Health Survey Report from 1971–1972 shows somewhat contradictory findings but did not measure the same type of acuity (Roberts 1967; Roberts, Lundford, 1977). The 1960–1962 survey measured binocular visual acuity; the later survey measured monocular acuity with the subjects' usual correction (glasses or contacts). By these standards U.S. Blacks have poorer acuity than Whites. Only 66% of Blacks reached the 20/20 level with correction, compared to 74% of Whites.

Because most of the information available on other groups in the survey is based on small samples, it is much less reliable. What can be teased out of the 1970–1971 survey is that the corrected visual acuity of Hispanic Americans is between that of Blacks and Whites (68.3% achieve 20/20). American Indians are comparable to Whites (78.2%), while Japanese Americans and Chinese Americans have lowest proportion achieving 20/20 visual acuity (50.3%) (Roberts, Lundford, 1977; Roberts, Rowland, 1978).

In general, visual acuity has improved over the past few decades for all groups. A rise in acuity also accompanies increases in education and family income (Roberts, Lundford, 1977; Roberts, Rowland, 1978). These differences in environment probably play a major role in acuity differences between various groups.

The poor visual acuity of Japanese and Chinese noted in the 1970–1971 National Health Survey is due mainly to myopia (nearsightedness). Oriental populations have long been noted to have a high incidence of myopia (Post, 1962). More recently Eskimos have begun to suffer from the same condition. Only in the past 40 years has myopia become common among these people. In fact, the newness of the condition, its general limitation to Eskimos under 40 years of age, and its increasing frequency in the younger age groups has led some to label it an epidemic (Morgan, Speakman, Grimshaw, 1975).

Previous to the Eskimo epidemic of myopia, it was thought that hunter-gatherers and other groups with a short history of civilization were less myopic than groups with a long history of civilization, because acute eyesight was necessary for survival in a hunting-gathering society. The increased number of myopes in older civilizations was thought to be the result of a relaxation of the need for good eyesight. The one-generation change from normal to myopic vision in a significant proportion of Eskimos negates this theory. It is now postulated that close indoor work may play a part in producing myopia (Morgan, Speakman, Grimshaw, 1975). Myopic Eskimos were shown to have attended school more regularly and for longer periods of time than their nonmyopic counterparts. The myopes had no higher proportion of myopic relatives than the normals, nor did they differ in height or weight. These comparisons were done to eliminate the effects of genetics or growth differences between myopes and nonmyopes. It seems that the tendency to become myopic has been present in the Eskimo population all along, but necessary conditions for the development of myopia were not present until recently. Animal studies have confirmed the role of visual experience in producing myopia (Wallman, Turkel, Trachtman, 1978).

Strabismus Racial differences in strabismus are not clear cut, but there is some evidence of a frequency difference in type of strabismus between Whites and Orientals (Ing, Pang, 1974). Two ophthalmologists in Hawaii found that the Whites in their practice had convergent strabismus more frequently than divergent strabismus, at a 3:2 ratio. Oriental patients had convergent less often than divergent strabismus, at a 1:3 ratio. There were over 200 cases for each race, and the results were statistically significant. Although this is only one study, the evidence for a racial difference is strong.

The National Health Survey also recorded strabismus. No differences were found in the frequency of all types of strabismus between Blacks and Whites (Roberts, Rowland, 1978). The numbers of Orientals were too small for use in analysis. The survey did not look at the types of strabismus; therefore no comparisons can be made with the Hawaiian study.

Epicanthic folds (extensions of the eyelids, originating at the inner canthus) may cause some Orientals and Native Americans to look cross-eyed when they are not. This pseudostrabismus can be distinguished from true strabismus by use of the corneal reflection test (DeMyer, 1980, pp 35–37).

Color blindness Color blindness varies by race, being higher in Whites and Orientals than in Blacks and American Indians (Post, 1962). It occurs at a frequency of 1–9% in men and rarely in women because it is an X-linked trait. It seems of little import to health care, except for a possible connection with hypertension. In 1975 a physician found a positive association between color blindness and hypertension in a large group of men being examined for military service (Morton, 1975). As far as I can determine, this study has not been replicated, so that the advice of the investigator, to follow color-blind individuals more closely for the possibility of hypertension, is not established.

Glaucoma An important eye disease, glaucoma, presents differently in different populations. The prevalence of blindness resulting from glaucoma is seven to eight times greater in Blacks than in Whites in the United States (Hiller, Sperduto, Drueger, 1982). Blacks show up for first treatment in a more advanced stage of disease than do Whites. They are also younger at first presentation and have higher intraocular pressures. It has been reported that Blacks have less pain with glaucoma than Whites, which could account for their delay in seeking treatment (Hansen, 1971).

Drug therapy for glaucoma is not as effective in Black patients as it is in Whites, though the reason has more to do with iris color than with race (Hiller, Sperduto, Drueger, 1982). Intraocular pressure also varies more by iris color than by race. Blacks with brown irises have the highest intraocular pressures, followed by Whites with brown irises, then whites with progressively lighter irises. It takes a larger dose of medication to produce a comparable reduction in pressure in the darker irises (Hiller, Sperduto, Drueger, 1982). Surgical therapy also seems to be less successful in Blacks than in Whites but for a different postulated reason: occlusion of the filter with scar tissue occurs more often in Blacks because of the increased tendency to keloid formation in Blacks (Hansen, 1971).

Other groups with a higher prevalence of glaucoma than Whites are Eskimos and Orientals (Alsbirk, 1974). These groups also have dark irises, which may increase their susceptibility to glaucoma.

Glaucoma examination should be carried out regularly on Blacks, Orientals, and Eskimos and at an earlier age than is recommended for Whites.

Hearing

Hearing acuity, like visual acuity, shows some differences by race, but again the differences are not clear-cut, and data are available for comparing Blacks and Whites only. The National Health Surveys of 1960–1961 and 1970–1971 gave a fairly consistent picture of Blacks having slightly better hearing at the low and high ends of the tested frequencies (250 and 6,000 hertz), while Whites had slightly better hearing at the middle frequencies (2,000 and 4,000 hertz) (Roberts, Bayliss, 1967; Roberts, 1972a; Roberts, Ahuja, 1975a; Rowland, 1980). The reason for this interesting difference is not known, but perhaps there is a relationship between these differences and the observed differences between Blacks and Whites in the incidence of otitis media. Blacks have the least otitis media, followed by Whites; Native Americans and Eskimos have the highest incidence of ear disease (Roberts, 1972b; Roberts, Ahuja, 1975b; Brody, Overfield, McAlister, 1965).

Taste

The only well-studied variation in taste by race is the ability to taste phenylthiocarbamide (PTC). This chemical and other similar chemicals are present in plants of the cabbage family (*Brassica* genus). Individuals who can taste PTC say it is bitter, and they also find the brassica group of vegetables bitter. In areas of the world with low iodine intakes, brassica plants are goitrogenic foods. It is thought that people who can taste the bitterness of such foods will avoid them and thus protect themselves. Populations vary in their frequency of individuals who can and cannot taste PTC. For instance, 30% of American Whites cannot taste PTC, but only 6% of American Blacks cannot taste it. Orientals and American Indians are intermediate (Greene, 1974). Because of the use of iodized salt in the United States, the incidence of preventable goiters is not high, but in other areas of the world the problem still exists.

A study of Ecuadoran children indicated that those who could taste PTC had better neurologic maturation scores than those who could not taste PTC (Greene, 1974). These children live in an endemic goiter area, and those who avoid the goitrogenic foods avoided putting stress on their thyroid glands; stressed thyroid glands could result in delays in neurologic development.

Smell

The sense of smell has lost importance in our sometimes overdeodorized and polluted world. Nevertheless, some individuals have retained or perhaps even trained their sense of smell to inform them of their surroundings. These people would probably be interested in the research showing that more Whites than Blacks can smell the odor of sweat, that more Blacks than Whites can smell the odor of musk (Whissell-Buechy, Amoore, 1973), that infants can detect their mothers' milk by smell (Russell, 1976), and that people can detect each other's sex by odor (Russell, 1976). But the topic has little to do with health care, so readers are left to sniff out their own odd facts about human responses to odor.

Touch

Pain is one aspect of the sense of touch that has been studied rather well from a cultural viewpoint but unfortunately has not been well studied biologically. Different cultures react differently to pain, but what needs further study is whether these differences are due to culture or biology (Pelz et al, 1981).

One large study of racial differences in pain tolerance exists (Woodrow et al, 1972). Over 40,000 subjects of various races were tested, using standardized pressure on the Achilles tendon (a measure of deep pain as opposed to cutaneous pain). The investigators found that Whites had the highest tolerance for deep pain, followed by Blacks, with Orientals having the lowest deep pain tolerance. There were also very distinct sex differences in all races. See Table 6.1. The differences among the races and sexes are all statistically significant.

The investigators also controlled for educational level in over 10,000 of their subjects 40–50 years of age; the racial differences remained. This finding gives some support to the notion that the differences in pain tolerance may have a biologic component that should be considered when treating individuals in pain.

The Woodrow study is supported by a smaller study that compared other groups. Jewish and other Mediterranean subjects have lower pain tolerance than northern Europeans—akin to that of Blacks (Chapman, Jones, 1944).

Common pain locations also seem to differ among groups. Lower back pain is one of the most common pain sites in British and American White subjects (Hall, Clarke, 1982; Pelz et al, 1981). In a Canadian Indian group, head pain was most common, and back pain was uncommon (Pelz et al, 1981).

NERVOUS SYSTEM DISEASE SUSCEPTIBILITY

Only two categories of disease are covered here: an example of an environmentally caused disease with population overtones, and congenital malformations of the nervous system.

Congenital malformations of the nervous system may differ among the races. Unfortunately, few studies include more than one racial group among their subjects, making comparisons between studies difficult, because of different diagnostic categories. One American study compared over 16,000 Black and 16,000 White births (Chung, Myrianthopoulos, Yoshizaki, 1967). Anencephaly was more than three times as frequent among Whites (1.4 per 1,000) as Blacks (0.4 per 1,000). No differences were found between the two races for meningomyelocele, meningocele, and encephalocele.

TABLE 6.1 Pain tolerance by race and sex

	Pressure on Achilles tendon (lb/sq in)		
Sex	White	Black	Oriental
Male	29.2	26.5	24.3
Female	16.1	15.2	14.4

Source: Adapted from Woodrow et al, 1972.

Another group of investigators compared their findings on Chinese births with the literature (Emanuel et al, 1972). They state that Chinese rates for anencephaly are intermediate (1.17 per 1,000) while the rates for spina bifida and hydrocephalus are low (0.16 per 1,000 and 0.35 per 1,000). A Japanese study (Neel, 1958) showed similar low rates for spina bifida and hydrocephalus, but it showed a low rather than an intermediate rate of anencephaly (0.60 per 1,000).

To confuse the issue, the findings of another American study, done on Whites, contradicted the other American study on anencephaly (Marden, Smith, McDonald, 1965).

The most that can be said about racial differences in congenital malformations of the nervous system is that indications of differences among populations exist, but there are insufficient data to make firm statements. The races do differ in other congenital deformities, as mentioned in the section on growth and development in chapter 3.

The evidence is clearer on the disease multiple sclerosis: it differentially affects various groups around the world. To some extent this disease, which is postulated to be environmentally caused, mimics racial variation in disease susceptibility. Multiple sclerosis occurs at a high frequency (30 cases per 100,000 population) in northern and central Europe, southern Canada, the northern United States, New Zealand, and southern Australia (Kurtzke, 1980). Medium frequency areas (5–20 cases per 100,000) include the southern United States and southern Europe. Low frequency areas (5 cases per 100,000) include Asia, Africa, Mexico, and Alaska.

Multiple sclerosis is not racially related except insofar as many Blacks originate in the low frequency areas of the southern United States and Africa, and many Orientals and Native Americans also originate in low frequency areas. This distribution may give the impression of racial differences in disease susceptibility. Data now accumulated demonstrate that, if a group migrates from a low to a high incidence area, within 15 years its incidence of multiple sclerosis rises to that of the new residence (Kurtzke, Huong, 1980). In the United States, the disease occurs less frequently in Orientals, Blacks, and American Indians, due more to their place of residence than their race.

BRAIN DEVELOPMENT

The growth and development section of chapter 3 noted that Black infants and children were advanced over Whites in motor development. This developmental maturity is reflected in the electroencephalogram (EEG). African, European, and Indian children under the age of 3 years were serially tested with EEGs over a period of 6–10 months (Pampiglione, 1965). In the occipital region, a well-developed sinusoidal rhythmic activity at 5–6 cycles per second appeared in African (Black) children by age 5–7 months, but in the European and Indian children this activity did not appear until 9 months. Also in the occipital area a definite alpha rhythmic activity at 8–9 cycles per second appeared at 18–24 months in African children, but in the European and Indian children it did not appear until 24–28 months of age. These differences in brain maturity parallel the differences seen between the races as measured by the Bayley and DDST tests.

It was also mentioned in chapter 3 that White children caught up with Black children by the time they entered school. This is reflected in a measure of intellectual maturity done as part of the National Health Survey. Using the Goodenough-Harris Drawing Test, children of several races between the ages of 6 and 11 years were tested for intellectual maturity (Harris, Roberts, 1972). There emerged few statistical differences in this measure of Black, White, Oriental, and Native American children, although Orientals and Native Americans tended to be slightly advanced over Whites, who were in turn slightly advanced over Blacks.

One other interesting racial difference in brain development is cerebral speech lateralization differences between Orientals, American Indians, and other races. It appears that Hopi, Navajo, and Japanese individuals process linguistic information in both hemispheres or have right cerebral hemisphere dominance (Scott et al, 1979). This finding is demonstrable in Navajos; when given a dichotic listening task, they have a left ear advantage (right cerebral hemisphere dominance) as opposed to Whites, who have the usual right ear advantage (left cerebral hemisphere dominance). Those who rehabilitate aphasia victims should be alert to possible racial differences in cerebral dominance.

MENTAL ILLNESS

There is no clear evidence of differences between races in mental illness rates. But abundant evidence shows that low socioeconomic status, low educational level, migration, divorce, separation, and being female all increase the risk of depression. The only biologic variable—sex difference—will be discussed in chapter 7. I had hoped to show that, when socioeconomic variables are controlled, symptoms of mental illness differ by group but the prevalence is similar among groups. What I found was that mental illness statistics are hard to decipher. The data for large populations usually report such things as use of mental health facilities or cost of treatment, not rates of disease. The few studies in which mental illness is counted use varying operational definitions.

A few studies seem to corroborate my thesis that mental illness rates are similar among groups. A large study of Blacks (382), Whites (2,638), and Mexican-Americans (417) done by the California State Department of Health found no differences among the three groups in emotional or mental illness (Roberts, 1980). A study of three rural towns in North Carolina showed that 24% of the people had definite or probable psychiatric disorders (Yahraes, 1973). And, depending on the study, 26–37% of Eskimos have psychiatric symptoms (Sanpath, 1975). Roughly a quarter to a third of a population seem to have psychiatric symptoms of some kind.

Summary

The sparse data presented in this chapter indicate that much further research is needed to establish differences, if any, among populations in characteristics of the nervous system. Health care providers also need to know more about glaucoma symptoms in various races. The areas of touch sensitivity and pain tolerance have barely been explored from a racial perspective. The known differences between races in maturation rates should be studied further. And cerebral speech lateralization studies are particularly needed by rehabilitation specialists and teachers.

REFERENCES

Alsbirk PH: Anterior chamber depth in Greenland Eskimos: II. Geographical and ethnic variation. *Acta Ophthalmol* 1974; 52(4):565–580.

Brody JA, Overfield T, McAlister R: Draining ears and deafness among Alaskan Eskimos. *Arch Otolaryngol* 1965; 81(1):29–33.

Chapman WP, Jones CM: Variation in cutaneous and visceral pain sensitivity in normal subjects. *J Clin Invest* 1944; 23:81–91.

Chung CS, Myrianthopoulos NC, Yoshizaki H: Racial and prenatal factors in major congenital malformations. *Am J Hum Genet* 1968; 20:44–60.

DeMyer W: *Technique of the Neurologic Examination*. McGraw-Hill, 1980.

Emanuel I et al: The incidence of congenital malformations in a Chinese population: The Taipei Collaborative Study. *Teratology* 1972; 5(2):159–170.

Greene LS: Physical growth and development, neurological maturation, and behavioral functioning in two Ecuadorian Andean communities in which goiter is endemic: II. Taste sensitivity and neurological maturation. *Am J Phys Anthropol* 1974; 41(1):139–151.

Hall W, Clarke IMC: Pain and laterality in a British pain clinic sample. *Pain* 1982; 14:63–66.

Hansen AC: Comparative ophthalmology: Black and White. *J of Natl Med Assoc* 1971; 63(6):445–449.

Harris DB, Roberts J: Intellectual maturity of children: Demographic and socioeconomic factors. *Vital Health Stat* 1972; 11(116):1–74.

Hiller R, Sperduto RD, Drueger DE: Race, iris pigmentation, and intraocular pressure. *Am J Epidemiol* 1982; 115(5):674–683.

Ing MR, Pang SWL: The racial distribution of strabismus—a statistical study. *Hawaii Med J* 1974; 33(1):22–23.

Kurtzke JF: Geographic distribution of multiple sclerosis: An update with special reference to Europe and the Mediterranean region. *Acta Neurol Scand* 1980; 62:65–80.

Kurtzke JF, Huong BQ: Multiple sclerosis in a migrant population: II. Half-Orientals immigrating in childhood. *Ann Neurol* 1980; 8(3):256–260.

Marden PM, Smith DW, McDonald MJ: Congenital anomalies in the newborn infant, including minor variations. *J Pediatr* 1965; 64:357–371.

Morgan RW, Speakman JS, Grimshaw SE: Inuit myopia: An environmentally induced "epidemic"? *Can Med Assoc J* 1975; 112(5):575–577.

Morton WE: Hypertension and color blindness in young men. *Arch Intern Med* 1975; 135(5):653–656.

Neel JV: A study of major congenital defects in Japanese infants. *Am J Hum Genet* 1958; 10:398–445.

Pampiglione G: Brain development and the EEG of normal children of various ethnical groups. *Br Med J* 1965; 1965(2):573–575.

Pelz M et al: A note on the occurrence of pain in psychiatric patients from a Canadian Indian and Inuit population. *Pain* 1981; 10(1):75–78.

Post RH: Population differences in visual acuity: A review with speculative notes on selection relaxation. *Eugen Q* 1962; 9(4):189–212.

Roberts J: Binocular visual acuity of adults. *Vital Health Stat* 1964; 11(3):1–27.

Roberts J: Binocular visual acuity of adults by region and selected demographic characteristics. *Vital Health Stat* 1967; 11(25):1–39.

Roberts J: Hearing levels of children by demographic and socioeconomic characteristics. *Vital Health Stat* 1972a; 11(111):1–47.

Roberts J: Hearing and related medical findings among children: Race, area, and socioeconomic differentials. *Vital Health Stat* 1972b; 11(122):1–33.

Roberts J, Ahuja EM: Hearing levels of youths 12–17 years. *Vital Health Stat* 1975a; 11(145):1–84.

Roberts J, Ahuja EM: Hearing sensitivity and related medical findings among youths 12–17 years. *Vital Health Stat* 1975b; 11(154):1–112.

Roberts J, Bayliss T: Hearing levels of adults by race, region, and area of residence. *Vital Health Stat* 1967; 11(26):1–29.

Roberts J, Lundford J: Monocular visual acuity of persons 4–74 years. *Vital Health Stat* 1977; 11(201):1–67.

Roberts J, Rowland M: Refraction status and motility defects of persons 4–74 years. *Vital Health Stat* 1978; 11(206):1–124.

Roberts RE: Prevalence of psychological distress among Mexican-Americans. *J Health Soc Behav* 1980; 21(6):134–145.

Rowland M: Basic data on hearing levels of adults 25–74 years. *Vital Health Stat* 1980; 11(215):1–49.

Russell M: Human olfactory communication. *Nature* 1976; 260:520–522.

Sanpath HM: Prevalence of psychiatric disorders in a southern Baffin Island Eskimo settlement. *Can Psychiatr Assoc J* 1975; 19(4):363–367.

Scott S et al: Cerebral speech lateralization in the native American Navajo. *Neuropsychologia* 1979; 17(1):89–92.

Wallman J, Turkel J, Trachtman J: Extreme myopia produced by modest change in early visual experience. *Science* 1978; 201(4362):29.

Whissell-Buechy D, Amoore JE: Odour-blindness to musk: Simple recessive inheritance. *Nature* 1973; 19(3):271–273.

Woodrow KM et al: Pain tolerance: Differences according to age, sex, and race. *Psychosom Med* 1972; 34(6):548–556.

Yahraes H: Studies of rural life and mental health. Pages 9–27 in: *The Mental Health of Rural America*. Segal J (editor). U.S. Department of Health, Education, and Welfare, 1973.

chapter 7

Sexual variation

Differences between the sexes are as evident as racial differences. The long fascination with and mythology about race are exceeded by the longer fascination with and greater mythology concerning sex. After all, differences between males and females were noticed way before early humans ever traveled far enough to find people who looked different from themselves.

The actual anatomic and hormonal differences between the sexes will not be covered here; these are well known and well studied. This chapter is concerned with other differences between males and females—reactions to common stressors, physical strength, physical performance, and brain differences. "In most respects sex differences in biological and pathological phenomena are questions of either more or less, but not of either/or" (Glucksmann, 1981, p 303). The chapter concludes with an explanation of why the observed differences occur.

REACTIONS TO STRESSORS

Multitudinous stressors impinge on humankind. Disease, malnutrition, and the physical and psychologic wear and tear of daily living are some common stressors. Reactions to them differ by sex.

Life Span and Health Maintenance

One general measure of stress resistance is life span. Women live longer than men. The effect of sex on life span is stronger than that of race. In 1980, the estimated life expectancy at birth in the United States, calculated for Whites and non-Whites of both sexes, showed that White females had the longest potential life span (78.1 years), followed by non-White females at 74.0 years. Next came White males at 70.5 years, and last non-White males at 65.3 years (National Center for Health Statistics, 1981). The death rate indicates this same advantage for females, regardless of race. It increases from a low of 516 per 100,000 for White women to a high of 1,300 per 100,000 for non-White men. Intermediate are non-White women at 825 per 100,000, and White men at 905 per 100,000 (Klebba, Maurer, Glass, 1973).

Another general measure of stress resistance is maintenance of health. Males generally fare better than females in this area. They experience fewer acute diseases than females, except before the age of 6, when females are healthier (Givens, 1979). Males spend fewer days in bed or on a restricted activity regime than do females; this holds for all ages. Between 17 and 64 years of age, males visit physicians less often than females do; but before age 17 and after age 64 the two sexes see physicians at comparable rates.

Hospitalization rates follow a similar pattern. Between ages 17 and 64, males are hospitalized less often than females. But before 17 and after 64 they are hospitalized slightly more often than females (Givens, 1979). While males generally are not hospitalized as often as females, they spend an average of two days longer in the hospital.

Injuries are the one acute condition that males incur more often than females. This holds for all ages except 64 years and older, when females experience more injuries (Givens, 1979). The majority of injuries in these older women occur in the home. Probably the injuries are broken bones due to falls.

Diseases

A few common disease syndromes that show sexual variation are diabetes, gallbladder disease, tooth decay, osteoarthritis, and coronary heart disease. These are not the only diseases that manifest sexual variation but they should suffice as illustrations. For additional information see Glucksmann (1981).

In the United States, maturity onset diabetes occurs at a ratio of 1.3 women for every man with the disease (Bauer, 1967). Gallbladder disease also exhibits a propensity for women; here the ratio is approximately 2 women for every man afflicted (Brown, Christensen, 1967; Erickson, 1972). Diabetes occurs more frequently in obese women than in women of normal weight. Obesity is also more frequent in women than in men. In the case of gallbladder disease, increased parity is associated with an increased risk of developing the disease (Comess, Bennett, Burch, 1967).

Females experience more tooth decay than males, and more women than men lose all their teeth (Burnham, 1974; Kelly, Harvey, 1979). The differences in tooth decay are not large (amounting to about two more decayed teeth), but the difference is consistent at all ages and in most groups that have been studied (Kelly, Harvey, 1979; Moller, Pindborg, Roed-Petersen, 1972). The effect of race is stronger than sex for tooth decay. Black males have the least tooth decay as measured by the decayed, missing, and filled (DMF) tooth score. Their DMF score is 8.9 teeth. White females have the highest DMF scores (14 teeth). White males at 12.9 DMF teeth and Black females at 10.3 DMF teeth are intermediate (Kelly, Harvey, 1979). The differences are probably due to the denser bones and teeth of Blacks compared with Whites (Trotter, Peterson, 1966).

Osteoarthritis, another disease almost as common as tooth decay, affects 85% of the population by the seventh decade of life (Roberts, Burch, 1966). Before age 45, men have more osteoarthritis than women, but after age 55 women have more of the joint disease than men. Men and women differ in their affected joints. Men usually have the disease in one or two pairs of joints, especially those that had strenuous use, while women tend to get multiple joint involvement (Jurmain, 1980).

When all the known predisposing factors are considered, coronary heart disease affects more men than women. After controlling for alcohol consumption, marital status, personality type, smoking, and working status, male rates are still higher than female rates (Talbott et al, 1977; Friedman, Dales, Ury, 1979; Haynes, Feinleib, 1980). Two factors that might contribute to the sex difference are higher plasma cortisol concentrations in men (possibly reflecting higher stress levels) and the presence of androgen receptors on the atrial and ventricular myocardial cells (indicating that sex hormones may directly affect the heart muscle) (Zumoff et al, 1974; McGill et al, 1980).

Psychologic Stress

Males and females react differently to stress as measured by the urinary excretion of catecholamines. Numerous studies have shown that women and girls under stress increase their catecholamine levels to a lesser degree than do men and boys (Frankenhaeuser, Dunne, Lundberg, 1976; Collins, Frankenhaeuser, 1978; Frankenhaeuser et al, 1978). As one author put it, women respond to stress in "a physiologically more economic way" (Frankenhaeuser et al, 1978, p 341). These data, coupled with the male propensity to suffer more from stress-sensitive coronary heart disease and the female longer life span, lend support to the notion that women handle psychologic stress more effectively than men.

Stress during Growth and Development

Females can withstand disease and malnutrition stress better than males during the growth and development period. Starting from birth, and probably from gestation onward, females survive better than males. The sex ratio at birth is 105 males born for every 100 females (National Center for Health Statistics, 1981). This is the secondary sex ratio; the primary sex ratio is the number of males per females at conception. There is still controversy about the primary sex ratio; estimates have ranged from 110:100 to 170:100 (McMillen, 1979). A good current estimate is 120:100 (McMillen, 1979). If the primary sex ratio is around 120:100 and the known secondary sex ratio is 105:100, many more males than females die in utero. The male disadvantage continues to be high in the neonatal period: neonatal deaths in one large study produced a sex ratio of 128:100 (Naeye et al, 1971).

Adverse conditions continue to affect boys more than girls. A study of poor Peruvian children indicated that, under the same conditions, more boys than girls died in the period from infancy through adolescence (Frisancho et al, 1973). By age 30, the sex ratio is even in the United States (Allison, 1964). After age 30, the sex ratio favors females, as males continue to die at younger ages than females.

The more dramatic sex differences in mortality that favor females are also reflected in nonfatal consequences of disease, malnutrition, and environmental stress. The hypoxic stress of high altitude decreases the birth weight and crown–heel length of male babies more than that of females (Haas et al, 1980). The multiple stresses of disease and malnutrition experienced by children living under marginal conditions stunt the growth of boys much more than of girls (Blanco et al, 1974). In fact, the differences in stature between males and females, in general, correlate strongly with environmental conditions. Where conditions are good, males are a few inches taller than females. Where conditions are marginal, the sex differences in height decrease or disappear (Hall, 1978). This ability of females to maintain close to normal growth under adverse conditions has been documented in numerous studies (Bielicki, Charzewski, 1977).

PHYSICAL PERFORMANCE AND STRENGTH

It is well known that, on the average, men are taller and more muscular than women. It is also obvious that women generally have more body fat than men. It is taken for granted that men are stronger than women. Most of these differences have been studied, and only a few surprises appear in the analyzed data.

The reasons for women being less strong than men and generally less able in physical performance tests are still being studied. Speculation on the causes of these discrepancies has centered on differences in muscle mass and type, on body size in general, and on differential use. As cultural expectations of appropriate physical behavior for males and females change, it is assumed that many of the differences in physical strength and performance will narrow or disappear. Sports literature is filled with accounts of how records in track and field, swimming, and like events are constantly being broken and how the women of today are achieving records that match the performances of men of a few decades ago (Jokl et al, 1966).

While men are generally stronger than women, they are also generally taller and heavier. Would the strength differences hold if size differences were taken into account? The answer is not a simple yes or no. Men are stronger in the upper extremities and trunk than women, but women are close to or perhaps equal in strength to men in their lower extremities, when body size is considered (Lauback, 1976; Hoffman, Stauffer, Jackson, 1979). A well-controlled study specifically measuring strength differences related to body size showed that men were almost twice as strong as women in the upper extremities, as measured by the amount of weight lifted in a bench press. They could lift an average of 152 pounds, compared to the women's average lift of 115 pounds (Hoffman, Stauffer, Jackson, 1979). In the leg press, women actually lifted an average of 29 pounds more than men when body size was controlled, but the difference was not statistically significant (710 pounds for women versus 681 pounds for men). These figures may be somewhat misleading, because they do not reflect the actual amount of weight lifted but rather the amount adjusted for body size. The actual amounts were 800 pounds for males versus 592 pounds for females in the leg press, and 178 pounds for males versus 89 pounds for females in the bench press. One possible conclusion from these findings is that, when body size is taken into account, men are indeed stronger than women in the upper extremities, but men and women have similar strength in their lower extremities.

Similarity of use is a fairly simple explanation for the similarity of strength in the lower extremities. Both sexes need to use their legs to get themselves from place to place. Use of transportation modes other than the feet is a fairly recent innovation from an evolutionary perspective.

If the sexes are probably equal in leg strength for their size, why do women not run as fast, jump as high, or do as well as men in the activities that rely on the legs? Aerobic power is a limiting factor. Women have less ability to transport and use oxygen maximally to support physical activity. Aerobic power is measured by Vo_2 max. This is the maximum oxygen uptake possible during a treadmill running test. Studies have shown that women have around 18–22% less aerobic power than men (Burke, 1977; Daniels et al, 1979). A difference exists even when physically fit young men and women are subjected to similar training regimens. The young women in two such studies had 18% less aerobic power than the young men with whom they trained (Burke, 1977; Daniels et al, 1979). The lessened aerobic capacity of women probably indicates that they reach oxygen consumption limitations before their potential muscle strength is exhausted. The reason women have less aerobic power probably relates to the fact that they have less hemoglobin (lower hematocrits), which would influence oxygen transport; small chest capacities might also be a factor (Burke, 1977; Protzman, 1979).

Not only are males more muscular than females, but also their muscles differ in several respects from those of females (Komi, Karlsson, 1978). During contraction, male muscles are neuromotorically more efficient. Males have more slow twitch muscle fibers than females, and there are pronounced differences in the contractile and glycolytic properties of their muscles compared with females. These muscle differences plus the aerobic power differences probably account for the strength differences between the sexes in the upper extremities.

BRAIN DIFFERENCES

Psychologists have produced some information that shows a difference in how each sex thinks and processes information. Biologists have produced some data that show differences between male and female brains and nervous systems. Physicians have noted differences in response to brain injury between the sexes. All this information adds up to the presence of some differences between the sexes in the brain and nervous system, but the data are still unclear regarding its extent. The differential response to brain injury and the biologic and psychologic findings generally support each other.

Since the psychologic discoveries are responsible to some extent for a recent surge of biologic and medical interest in the topic of sex differences, it is necessary to outline these findings as a base on which to build and view the biologic findings. Psychologists are almost unanimous in agreeing that male brains are organized more asymmetrically for verbal and nonverbal functions than are female brains (McGlone, 1980). Language and "the execution of learned manual activities" are mainly functions of the left cerebral hemisphere, while the right hemisphere is concerned more with "perception, construction, and recall of stimuli that are difficult to verbalize" (McGlone, 1980, p 215). For females the brain is less specialized for these functions than is the brain of males.

The lessened specialization of the female brain compared to the male brain accounts for the discrepancies found in patients with right or left cerebral damage (McGlone, 1980; Inglis, Lawson, 1981). Left brain damage produces impairment in verbal tests compared to nonverbal tests in male patients, but not in female patients. Right brain damage produces impairment in nonverbal tests compared to verbal tests in males, but again not in females. Females with left cerebral damage experience a decline in both verbal and nonverbal tests, but females with right cerebral damage do not show a decline on either test.

Other pieces of evidence of a functional brain difference by sex are the rate and severity of severe speech disorders (aphasia). Three to five times more men than women are afflicted with aphasia (McGlone, 1980; Sasanuma, 1980). In a study of 100 aphasia patients (78 males and 22 females), 14 females and 21 males were classified as having mild aphasia, 5 females and 23 males as moderate, and 3 females and 34 males as severe.

Anatomic brain differences between the sexes have been found in the ventricular system and the corpus callosum (Gyldensted, 1977; De Lacoste-Utamsing, Holloway, 1982). No interpretation was suggested for the ventricular system differences, but the investigators postulated that the females' more bulbous and larger posterior corpus callosum (splenium) is used for interhemispheric connection—its enlargement indicates more connecting fibers. The authors cite evidence from animal studies to support their hypothesis, along with evidence regarding the females' more symmetrical brain function.

Symmetry differences in brain function are involved in the hotly debated issues of whether there are biologic or sociocultural causes for the sex differences found in the mathematic and verbal portions of intelligence test results. Arguments for the sex differences being biologic are based on studies showing early brain function asymmetry in boys, overrepresentation of boys among mathematically gifted 12- and 13-year-olds, the effect of testosterone on brain function, and the association of maleness with several disparate brain functions and syndromes.

In the case of early brain asymmetry, boys as young as 6 years show evidence of asymmetrical brain function, whereas girls show symmetrical brain function up to 13 years of age (Witelson, 1976). The tests used to determine brain hemisphere specialization were probably culture-free tests, consisting of tasks that children (or adults) would never do in the course of ordinary life.

Data from almost 50,000 seventh and eighth grade students taking the Scholastic Aptitude Test (SAT) show that boys consistently score higher than girls on the mathematics section and that at the high-scoring end boys outnumber girls 13 to 1 (Benbow, Stanley, 1980; Benbow, Stanley, 1983). On the verbal portion of these tests boys and girls are equally represented. The authors of this study found no support for the sociocultural hypotheses that boys are socially reinforced toward mathematics and that boys have taken more mathematics courses than girls.

It has recently been observed that brain asymmetry—specifically right hemisphere specialization—is linked with left-handedness, language disorders, several autoimmune diseases, autism, stuttering, migraine headaches, and myopia (Marx, 1982; Kolata, 1983). This seemingly haphazard collection of conditions is linked, according to its discoverer, through the effect of testosterone on the brain and immune system. The authors of the large study of seventh and eighth graders reanalyzed their data to see if they could substantiate the connection between right hemisphere specialization and the various conditions. Of their 280 mathematically gifted children (260 boys and 20 girls), 20% were left-handed, 60% had some form of allergy or asthma, and 70% were myopic. Information was not obtainable on the other conditions. Boys and girls whose test scores were at the lower end, that is those with close to average test scores, had average amounts of left-handedness, allergies, and nearsightedness.

Another study links testosterone to spatial ability, a right hemisphere function. Investigators compared 19 men born with idiopathic hypogonadotropic hypogonadism (profound androgen deficiency) with 19 normal men and 5 men with acquired hypergonadotropic hypogonadism that developed after puberty (Hier, Crowley, 1982). The men with the congenital androgen deficiency scored significantly lower on spatial ability tests than the other two groups of men. But the verbal ability test scores were similar in all three groups. This supports the notion that testosterone exerts an influence on the brain before or at puberty.

The evidence for a sex difference in cerebral hemisphere function is substantial; the evidence on whether the differences are due to biologic or sociocultural causes is less robust. But the anatomic and biochemical evidence cited above prompts me to favor a biologic explanation.

One other sex difference in the brain and nervous system that is pertinent to health care workers is a difference in pain tolerance. Men can tolerate more pain than women. This has been demonstrated in a large study of 41,119 subjects from three races, ranging in age from 20 to 70 years (Woodrow et al, 1972). The sex difference held at all ages and within the three races. The pain measure was Achilles tendon pressure, a measure of deep pain, as opposed to superficial or cutaneous pain. The authors chose the deep pain measure because it has more clinical relevance. It is important to note that pain tolerance, not pain threshold, was measured. Pain medication is given "more on the basis of intolerance of pain and discomfort than on pain recognition" (Woodrow et al, p 552).

SEX DIFFERENCES—AN EXPLANATION

The previously mentioned sex differences are in large part due to the influence of the sex chromosomes. Of course, some sociocultural influences impinge, but biology is the more important influence. Females have a form of hybrid vigor because they have the benefit of two X chromosomes. Males must get along with only one X chromosome.

Random inactivation of one of the female's two X chromosomes occurs early in embryonic development; this results in females and males having a similar amount of genetic material. While males have only one X chromosome in all their cells, females have one or the other of their two X chromosomes in the cells throughout their bodies. Effectively, males are homozygous at their genetic loci on the X chromosomes, whereas females may be heterozygous or homozygous at their X chromosome loci, depending on which X chromosome inactivates in a particular cell and what they inherited from either parent.

There is evidence that portions of the short arm of the X chromosome escape inactivation (Gordon, Ruddle, 1981). The short arm contains the genes necessary for the maintenance of fertility. The portions that escape inactivation balance the genes on the Y chromosome, which assures that males and females have equal amounts of genetic material.

Hybrid vigor or heterozygote advantage is a reasonable explanation for the preferential survival of females in utero and throughout life. The wider choice of response at the genetic loci on the X chromosome could account for the differential survival rates between the sexes and the ability of female children to maintain close to normal growth patterns under adverse conditions in contrast to the stunting of growth seen in male children.

Another property of the X and Y chromosomes may explain the high sex ratio at conception (primary sex ratio). The Y chromosome is smaller than the X chromosome. A sperm cell containing a Y chromosome probably weighs less than one containing an X chromosome. Being lighter, a Y sperm cell should be capable of faster movement than an X sperm cell (Meyer, Diamond, Merz, 1968). Consequently it should have a better chance of reaching and fertilizing the ovum, thus accounting for the sex ratio at conception in favor of males.

The sex ratio at birth (secondary sex ratio) depends on survival of the fetus during the gestation period. More males are spontaneously aborted than females (Meyer, Diamond, Merz, 1968). There is also documentation that countries with a higher sex ratio at birth than the United States have lower infant mortality rates (Chase, 1967). Both of these factors point to the superior survival of female fetuses, another example of hybrid vigor.

The correlation of low infant mortality rates with high sex ratios at birth suggests that, under good conditions, more males will survive. Another example of this is seen in countries with better health conditions than those in the United States. In such countries the difference between female and male life expectancies is not as great as it is in this country. United States females experience similar life expectancies to females from most developed countries, while United States males fare less well (Tomasson, 1976). This results in the United States having a large sex difference in life expectancy, compared with many other developed countries. In Sweden, the sex difference in life expectancy is the smallest in the world. Sweden also has the lowest male mortality rate of any country. This provides further evidence for males having less ability to respond to adverse conditions than females. Where health conditions are excellent, males survive almost as long as females.

Some have argued that the sex difference in mortality is primarily due to sociocultural factors. But one telling argument against this position is that at all age levels and in most countries the sex differential remains. A longitudinal study of 6,928 adults indicated that biologic differences were more to blame for the differential survival of females than such behavioral risk factors as smoking, alcohol consumption, disability, physical inactivity, lack of marriage, and dissatisfaction with life (Wingard, 1982).

Even though women live longer, men seem to stay healthier than women between the ages of 17 and 64 years. Explaining this discrepancy is not easy. Perhaps the increased physical strength of men, their denser bones, and their lessened sensitivity to pain help to protect them from some of the physical stresses that disable women. But an argument could be made that women give in to illness and that men continue to maintain their daily pattern despite illness. The increased length of stay for men when they enter hospitals tends to support this argument—they hold out longer, thus getting sicker.

Summary

Like race differences, sex differences have often been overlooked or explained away as resulting from sociocultural influences. In the last decade or two, the notion has developed that once social, cultural, and economic conditions were equalized most differences between the sexes would disappear. This chapter contains data indicating that some of the biologically based differences will not change—physical strength, physical performance, and brain differences—and that females generally respond better to stress than males do.

REFERENCES

Allison AC: Polymorphism and natural selection in human populations. *Cold Spring Harbor Symp Quant Biol* 1964; 29:137–149.

Bauer ML: Characteristics of persons with diabetes. *Vital Health Stat* 1967; 10(40):1–44.

Benbow CP, Stanley JC: Sex differences in mathematical ability: Fact or artifact? *Science* 1980; 210:1262–1264.

Benbow CP, Stanley JC: Sex differences in mathematical reasoning ability: More facts. *Science* 1983; 222:1029–1031.

Bielicki T, Charzewski J: Sex differences in the magnitude of statural gains of offspring over parents. *Hum Biol* 1977; 49(3):265–277.

Blanco RA et al: Height, weight, and lines of arrested growth in young Guatemalan children. *Am J Phys Anthropol* 1974; 40(1):39–48.

Brown JE, Christensen C: Biliary tract disease among the Navajos. *JAMA* 1967; 202(11):138–140.

Burke EJ: Physiological effects of similar training programs in males and females. *Res Q Am Alliance Health Phys Educ Recreat* 48(3):510–517.

Burnham CE: Edentulous persons. *Vital Health Stat* 1974; 10(89):1–31.

Chase HC: International comparison of perinatal and infant mortality. *Vital Health Stat* 1967; 3(6):1–97.

Collins A, Frankenhaeuser M: Stress responses in male and female engineering students. *J Hum Stress* (June) 1978; 4(2):43–48.

Comess LJ, Bennett PH, Burch TA: Clinical gallbladder disease in Pima Indians: Its high prevalence in contrast to Framingham, Massachusetts. *N Engl J Med* 1967; 277(17):894–898.

Daniels WL et al: Physiological effects of a military training program on male and female cadets. *Aviat Space Environ Med* 1979; 50(6):562–566.

De Lacoste-Utamsing C, Holloway RL: Sexual dimorphism in the human corpus callosum. *Science* 1982; 216:1431–1432.

Erickson WG: Cholecystectomy and the American Indian. *Rocky Mt Med J* (July) 1972; 69:33–35.

Frankenhaeuser M, Dunne E, Lundberg U: Sex differences in sympathetic-adrenal medullary reactions induced by different stressors. *Psychopharmacology* (1976); 47:1–5.

Frankenhaeuser M et al: Sex differences in psychoneuroendocrine reactions to examination stress. *Psychosom Med* 1978; 40(4):334–343.

Friedman GD, Dales LG, Ury HK: Mortality in middle-aged smokers and nonsmokers. *N Engl J Med* 1979; 300(5):215–217.

Frisancho AR et al: Adaptive significance of small body size under poor socio-economic conditions in southern Peru. *Am J Phys Anthropol* 1973; 39(2):255–262.

Givens JD: Current estimates from the health interview survey. *Vital Health Stat* 1979; 10(130):1–98.

Glucksmann A: *Sexual Dimorphism in Human and Mammalian Biology and Pathology*. Academic Press, 1981.

Gordon JW, Ruddle FH: Mammalian gonadal determination and gametogenesis. *Science* 1981; 211:1265–1271.

Gyldensted C: Measurements of the normal ventricular system and hemispheric sulci of 100 adults with computed tomography. *Neuroradiology* 1977; 14:183–192.

Haas JD et al: Altitude, ethnic, and sex difference in birthweight and length in Bolivia. *Hum Biol* 1980; 52(3):459–478.

Hall RL: Sexual dimorphism for size in seven nineteenth century northwest coast populations. *Hum Biol* 1978; 50(2):159–172.

Haynes SG, Feinleib M: Women, work, and coronary heart disease: Prospective findings from the Framingham Heart Study. *Am J Public Health* 1980; 70(2):133–141.

Hier DB, Crowley WF: Spatial ability in androgen-deficient men. *N Engl J Med* 1982; 306(20): 1202–1205.

Hoffman T, Stauffer RW, Jackson AS: Sex differences in strength. *Women in Sports* 1979; 7(4):265–267.

Inglis J, Lawson JS: Sex differences in the effects of unilateral brain damage on intelligence. *Science* 1982; 212:693–695.

Jokl E et al: Sports medicine. *Ann NY Acad Sci* 1966; 134(2):908–937.

Jurmain RD: The pattern of involvement of appendicular degenerative joint disease. *Am J Phys Anthropol* 1980; 53(1):143–150.

Kelly JE, Harvey CR: Basic dental examination findings of persons 1–74 years. *Vital Health Stat* 1979; 11(214):1–33.

Klebba AJ, Maurer JD, Glass EJ: Mortality trends: Age, color, and sex, United States, 1950–69. *Vital Health Stat* 1973; 20(15):1–40.

Kolata G: Math genius may have hormonal basis. *Science* 1983; 222:1312.

Komi PV, Karlsson J: Skeletal muscle fibre types, enzyme activities and physical performance in young males and females. *Acta Physiol Scand* 1978; 103:210–218.

Lauback LL: Comparative muscular strength of men and women: A review of the literature. *Aviat Space Environ Med* 1976; 47(5):534–542.

Marx JL: Autoimmunity in left-handers. *Science* 1982; 217:141–144.

McGill HC et al: The heart is a target organ for androgen. *Science* 1980; 207:775–777.

McGlone J: Sex differences in human brain asymmetry: A critical survey. *Behav Brain Sci* 1980; 3:215–263.

McMillen MM: Differential mortality by sex in fetal and neonatal deaths. *Science* 1979; 204:89–90.

Meyer MB, Diamond EL, Merz T: Sex ratio of children born to mothers who had been exposed to x-rays in utero. *Johns Hopkins Med J* 1968; 123(3):123–127.

Moller IJ, Pindborg JJ, Roed-Petersen B: The prevalence of dental caries, enamel opacities, and enamel hypoplasia in Ugandans. *Arch Oral Biol* 1972; 17:9–22.

Naeye RL et al: Neonatal mortality, the male disadvantage. *Pediatrics* 1971; 48(6):902–906.

National Center for Health Statistics: Annual summary of births, deaths, marriages, and divorces: United States, 1980. *Monthly Vital Stat Rep* 1981; 29(13):1–31.

National Center for Health Statistics: Advance report of final natality statistics, 1979. *Monthly Vital Stat Rep* 1981; 30(6):1–27.

Protzman RR: Physiologic performance of women compared to men. *Am J Sports Med* 1979; 7(3):191–194.

Roberts J, Burch TA: Osteoarthritis prevalence in adults by age, sex, race, and geographic area. *Vital Health Stat* 1966; 11(15):1–27.

Sasanuma S: Do Japanese show sex differences in brain asymmetry? Supplementary findings. *Behav Brain Sci* 1980; 3:247–248.

Talbott E et al: Biologic and psychosocial risk factors of sudden death from coronary disease in White women. *Am J Cardiol* 1977; 39:858–864.

Tomasson RF: The mortality of Swedish and U.S. White males: A comparison of experience, 1969–71. *Am J Public Health* 1976; 66(10):968–973.

Trotter M, Peterson RR: Some variable factors in the adult skeleton. *Ann NY Acad Sci* 1966; 134(2):841–845.

Wingard DL: The sex differential in mortality rates: Demographic and behavioral factors. *Am J Epidemiol* 1982; 115(2):205–216.

Witelson SF: Sex and the single hemisphere: Specialization of the right hemisphere for spatial processing. *Science* 1976; 193:425–427.

Woodrow KM et al: Pain tolerance: Differences according to age, sex, and race. *Psychosom Med* 1972; 34(6):548–556.

Zumoff B et al: The sex difference in plasma cortisol concentration in man. *J Clin Endocrinol Metab* 1974; 39:805–808.

chapter 8

Biologic variation in Sioux Indians

by BARBARA CRISS, RN, MS, CNM

After being introduced to the concept of biologic variation, Barbara Criss, a nurse-midwife, put together the following information. The specific data plus her 10 years of work with the Sioux resulted in a manuscript containing vital information for health care providers who work with Sioux Indians specifically and Native American groups generally. This chapter is an example of what should be done by other health care providers who work with specific racial groups.

The Pine Ridge Reservation on the high plains is the second largest reservation in the 48 contiguous states (Maynard, Twiss, 1970). It is the home of the Oglala Sioux, who are members of the Teton Dakota division of the Siouan family. The seven Teton Dakota tribes all speak the Lakota dialect of the Hokan-Siouan language group (Wissler, 1940).

The Sioux are classified as American Mongoloids, as are the other American Indian tribes. It is presumed that the ancestors of these tribes migrated at various times from Asia across the land-bridge of Alaska (Griffin, 1968). Regardless of the time of migration, all are descended from the Asian Mongoloids (Brues, 1977).

The Sioux are believed to have subsequently migrated westward from an earlier location in the southeastern United States. They had reached Minnesota by the 16th century and the region of the upper Missouri by the 18th century (Hyde, 1976). Teton Dakota history is well documented from the 19th century and demonstrates the conflict between Native Americans and Europeans from the times of earliest contact until the present (Spencer, Jennings, 1965). The advent of the horse in the middle of the 18th century not only provided the Sioux with an efficient means of access to game, particularly buffalo, but also gave them greater security and an environment where actual overabundance and wealth were possible (Hassrick, 1964). Yet, the Sioux period of dominance of the northern plains, like others in history, was brief; the tribes developed, flourished, and lost ascendancy in little more than a century. The Sioux epoch of wealth and abundance ended with the arrival of the Whites, who initially brought disease and later, as a result of the discovery of gold in the Black Hills, took over land that had been given to the Sioux by the Treaty of 1868 (Josephy, 1968; Brandon, 1974).

The Oglala Sioux were assigned to the Pine Ridge, South Dakota, Reservation by a new bill in 1876 and finally settled there in 1878. The allotment system initiated by the Dawes Act in 1887 further reduced the land base of the Sioux. By the mid-1920s the land base of the Native Americans had been reduced from 140 million acres to 32 million acres (Jorgensen, 1978). This and other treaty violations led to an early and persistent Indian distrust of the government.

A Pine Ridge research bulletin reflects the problems that exist today on reservations:

> In general, Indian Reservation communities experience more poverty, less employment and restricted opportunities for personal advancement when compared to non-reservation communities .... Lack of educational attainment also limits employment opportunity, and the people of the Pine Ridge reservation face problems of housing, sanitation and transportation which make big city ghettos seem privileged in comparison [Stuart, 1969, p 77].

GENERAL DEMOGRAPHIC HEALTH FEATURES

The age distribution of the Native American population in the United States differs radically from the general United States population. The 1970 census gives the median age for Native Americans as 20.5 years and the median age for White Americans as 28.9 years. Population pyramids further clarify the sex-age distribution for the White and Indian populations of Pine Ridge in 1968 (see Figure 8.1). The Indian population pyramid is almost a true pyramid; this is often indicative of an economically underdeveloped population with high birth and mortality rates. The non-Indian population pyramid shows a larger percentage of persons in the productive middle (20–49) age groups, while the Indian pyramid shows that this most productive group is relatively small to carry a heavy portion of the responsibility for support of the remainder of the population (Maynard, Twiss, 1970).

The five leading causes of death in the Aberdeen Area, of which Pine Ridge is a part, and for the Indian Health Service in general, are accidents,

FIGURE 8.1 Indian and non-Indian population pyramids

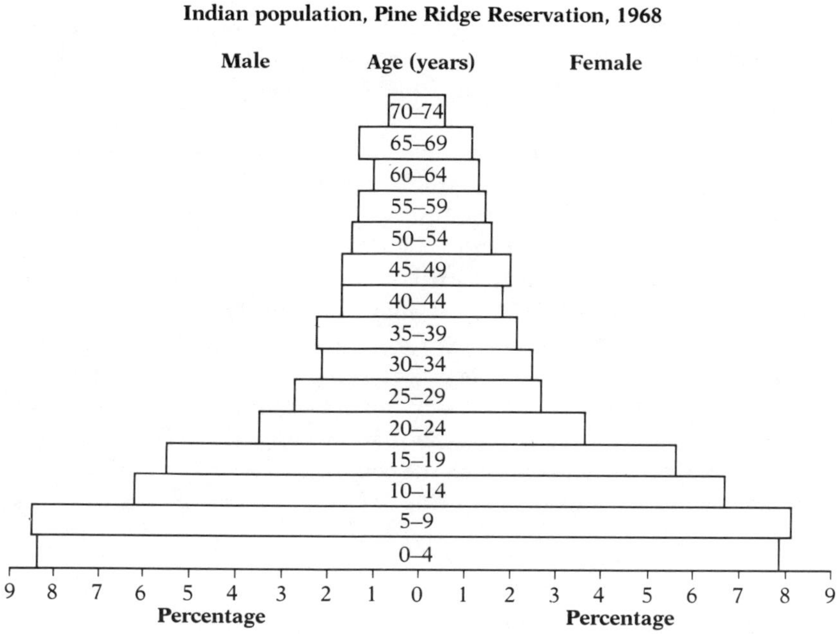

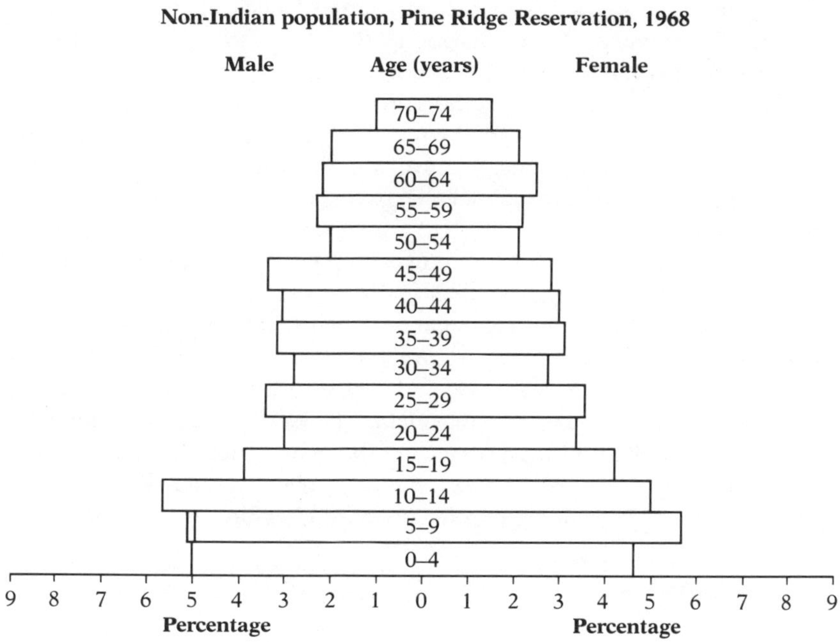

Source: Maynard, Twiss, 1970, pp 27–28.

diseases of the heart, malignant neoplasms, cirrhosis of the liver, influenza, and pneumonia. The live birth rate per 1,000 population is 38 for Native Americans and 18 for Whites. The Pine Ridge population increased from 10,146 in 1972 to 11,458 in 1977. Infant deaths declined from 51.7 per 1,000 births in 1972 to 26.0 in 1977, and live births declined from 40.0 per 1,000 in 1972 to 33.5 in 1977. The comparable infant mortality rates for all races during the years 1972–1977 in the United States declined from 18.5 to 16.5. The neonatal mortality rates by race for 1971 indicate that only Black Americans have higher rates than Indians. The mean family income is lower for American Indians than for all other ethnic groups. There has been an increase in births to mothers 15–19 years old in the Aberdeen Area, from 490 in 1972 to 549 in 1977. While this reflects a nationwide trend of increasing teenage pregnancies, it is particularly important for health care providers to notice in a population that has so many young members. Special programs need to be designed through the cooperative efforts of health, education, and tribal personnel to provide information and services to these young people.

HEALTH CARE FACILITIES AND STAFF

In 1955 the provision of health care for Native Americans was transferred from the Department of the Interior to the United States Public Health Service (USPHS) (Chaudhuri, Chaudhuri, 1973). For some time almost all professional staff members were personnel of the Commissioned Corps of the USPHS. This is no longer so; the majority of Indian Health Service employees are now civil servants. A move is under way for tribes to manage their hospitals and more personnel are coming from the ranks of tribal employees. Today the USPHS in Pine Ridge operates an accredited 58-bed hospital. The staff includes 2 nurse-midwives, 7 general practitioners, and 1 surgeon. There are 3 dentists and a 40-person nursing staff. The Pine Ridge Service Unit has 100 employees, many of whom staff the outlying district clinics that bring health care closer to the people. Clinics are situated from 35 to 100 miles away from the hospital. Transportation is a problem on a reservation that is the size of Connecticut. The work of health personnel is enormously benefited by the presence of indigenous health workers called community health representatives. They are tribal employees who focus their activities on prevention, chronic disease follow-up, and patient liaison.

POPULATION VARIATIONS PERTINENT TO HEALTH CARE

Lactose Intolerance

The majority of Native Americans cannot tolerate milk because of a deficiency of lactose enzyme. Some 60–80% of adult Indians exhibit intolerance, evidenced by abdominal distention, cramping, and diarrhea, when milk is ingested (Hongladarom, Russell, 1976; Bayless et al, 1975; Duncan, Scott, 1972; Caskey et al, 1977; Newcomber et al, 1977).

Normal children from all racial backgrounds tolerate milk until 3 to 5 years of age. Genetic factors determine a person's tolerance to lactose after age 5. Because tolerance is low worldwide, it is unwise to promote milk drinking after age 5 (Lebenthal, 1979).

Diarrheal diseases remain common among American Indians. This insult of mucosal injury to the small intestine often precedes the development of lactose intolerance. Infants, particularly up to age 18 months, need a careful reintroduction of lactose after a diarrheal episode, especially during the first months following injury (Lebenthal, 1979).

Implications Health care providers need to be aware that the process of converting milk to powdered or evaporated form increases the percentage of lactose (Hongladarom, Russell, 1976). Thus, two commonly distributed products in the Commodity Food Program actually contain higher amounts of lactose than the unprocessed food item.

The Women, Infants, and Children (WIC) Program encourages the utilization of milk. Eligible recipients of the WIC Program often come from ethnic populations that are lactose intolerant.

Yogurt, aged cheese, and buttermilk may be suggested as calcium sources for children from Native American families with less traditional food patterns. Those with traditional dietary habits accept the notion of including bones in soup, eating the softened ends of poultry bones, and cracking roasted meat bones for marrow. Canned fish, especially the bony varieties, can be recommended. Canned salmon is a delicacy among plains tribes, and sardines are often included in the diet. All these bony fish are good calcium sources.

Other sources of calcium are nuts and the highly popular peanut butter. The latter is distributed in commodity programs, and its price is not prohibitive. Dietary personnel can be urged to prepare peanut butter cookies and make peanut butter sandwiches for all age groups from Head Start programs through high school to nursing home residents. The elderly often have unmet calcium needs. Their sense of contribution can be fostered if they are asked to add their homemade native fruit jellies and jams to the peanut butter sandwiches.

Infant cereals are excellent sources of calcium. The cereals may be purchased by WIC voucher, thus increasing their accessibility to economically compromised groups. These tasty cereals can also be sprinkled on dry cereals or added to cooked cereals for the rest of the family. But people need to be taught to become label readers because, if the source of the increased calcium in a cereal is added dry milk, it is a poor choice for lactose-intolerant populations.

It is imperative to avoid diarrhea because of the sequelae of mucosal damage. Preventive teaching of handwashing, water boiling, sanitary formula preparation, and food storage technique should be part of postpartum teaching programs. Public health nurses and community workers need to be alert for children with early signs of diarrhea and to treat them appropriately. They should also reinforce diarrhea prevention in the home setting.

The best preventive measure for infant diarrhea is breast-feeding. There is less severe diarrheal disease among breast-fed infants, and many areas of potential contamination of the food supply and food storage problems can be avoided. Breast-feeding offers some protective factor against diarrhea.

The distribution of milk in the Commodity Food Program must be reevaluated. Routine distribution of milk at meals to schoolchildren, hospitalized patients, and the elderly on reservations should be questioned.

Yogurt is a good alternative to meet calcium requirements. It has the consistency of a much-liked indigenous food item, wojapi—a Sioux Indian pudding with berries. Blueberry is the favored pudding flavor, and this flavor could be introduced in trial studies. Simple procedures for making yogurt could be introduced in school.

Health personnel and school personnel need specific education regarding lactose intolerance. They should not urge children or adults to drink milk. See chapter 5 for more information on nutritional counseling. Community education programs on the symptoms of milk intolerance, the importance of preventing diarrhea, and the importance of breast-feeding need to take place on an ongoing basis.

Hereditary Polymorphic Light Eruption

Indians who dwell in the central plains of the Americas have a hereditary form of light-caused dermatitis characterized by an early onset. This dermatitis is the result of a genetic disturbance that appears to be autosomal dominant (Fusaro, Johnson, 1980). The photosensitivity typically has a long course with a tendency to persist in mild form even in winter (Birt, 1968). See chapter 4 for a description of the lesions.

Before acculturation, it was common for Native American peoples to wear hats, high collars, and clothing that covered their arms and legs, so that these areas were not exposed to the sun. The old-time cradle boards also had sun shields. With the adoption of less protective clothing they are more likely to develop photosensitive skin eruptions. This is especially the case in Indian women and children (Birt, 1968). The skin eruptions seem to be rare in the general population but frequent among Indians (Everett et al, 1961). The presence of solar sensitivity is unexpected because of the relatively dark color of the exposed areas of the skin (Schenck, 1960). By age 5, 35% of those who will ever have the disease have developed light eruption; by age 10, 70% have developed lesions; and by age 16, 75% show lesions (Birt, Davis, 1975). As many as 75% of patients in one study had a positive family history of photosensitivity. Thus far, hereditary polymorphic light eruption has not been found in the Indians of the Pacific Northwest or in Eskimos (Birt, Davis, 1975).

Implications The important etiologic agent in hereditary polymorphic light eruption is sunlight. Health teaching should include advice on wearing long-sleeved clothing and wide-brimmed hats. Further, pediatric staff members should consider that children admitted with impetigo may have been exposed to excessive sunlight and may have photodermatitis secondarily infected with impetigo. They should counsel mothers of such children to provide meticulous skin care, to avoid other children with infections, and to dress their children appropriately for sun protection.

Otitis Media

A 7.9% incidence of otitis media has been documented among the Navajo school population. This is fifteen times the 0.5% incidence among Pittsburgh school children (Johnson, 1967). Chronic otitis media, previous tympanoplasty or mastoidectomy, healing necrotizing otitis media, and adhesive otitis media collectively are present in 15.9% of the Navajo school-age population (Johnson, 1967).

The association of chronic otitis media with bottle-feeding was shown in a study of Inuit children. In the same families, older children who had been breast-fed had remarkably little otitis media compared with younger bottle-fed siblings (Timmermans, Gerson, 1980).

Among the Eskimos, otitis media has been a leading cause of morbidity. The first year of life is the year of highest risk for the first episode of draining ears (Reed, Struve, Maynard, 1967). Approximately one third of the village Eskimos are found to have chronic otitis media within the first two years of life (Brody, Overfield, McAlister, 1965).

Although recent incidence rates among the Sioux are not available, both anecdotal evidence and similarities in disease patterns suggest that otitis media is a prevalent disease among the Sioux. Recently two authors stated that "today in the Upper Missouri River Basin... otitis media is the most common infection, after the common cold" (Gregg, Steele, 1982).

Implications The progression of chronic otitis media to hearing loss makes it a disease worthy of vigorous treatment and community education efforts. Because onset often occurs prior to age 2, mothers need to be urged to bring children in for examination, particularly when an upper respiratory infection seems to be following an untoward course.

A good opportunity to initiate a thorough teaching program is available during the postpartum period of hospitalization. Since draining ears may be common, it is imperative that health teaching emphasize that they are abnormal and require medical attention. This teaching needs reinforcement in the home by health workers and community health nurses.

The importance of breast-feeding for the prevention of ear disease should be emphasized in community health education programs and prenatal clinics. Teaching should not focus solely on the primiparous patient. Many patients decide on breast-feeding for the first time during subsequent pregnancies, often as the result of observing healthier breast-fed infants. Mothers who can compare their breast-fed children with their bottle-fed children often report the former to be healthier. These mothers can be encouraged to become effective educators by sharing this information with their neighbors.

Dry Earwax

Earwax has been classified into two types, as described in chapter 3. It functions to prevent dryness of the ear canal but has no antifungal or antibacterial properties. Neither type has been associated with ear disease. What has been shown is a racial pattern in the occurrence of dry versus wet earwax (Matsunaga, 1962).

Examination of earwax type among Native Americans reveals a similarity to the Japanese, mostly dry earwax. The later-arriving Navajo show a pattern of dry cerumen most like the Oriental races; the Siouan tribes, who migrated earlier and thus were exposed for a greater length of time to population reductions from disease as well as admixture with Whites, are least like the Oriental races (Petrakis, Molohon, Tepper, 1967). The Sioux have an intermediate frequency of dry earwax, less than Navajos and Orientals but more than Whites.

Implications It is important to inform parents of the presence of wet earwax and to provide anticipatory guidance that these children will often require irrigations of the external auditory canal to make visualization of the tympanic membrane possible. The parents need to make this information available to subsequent examiners. As a matter of interest, it can be shared with parents that adults and children with dry wax often have no axillary odor and can avoid the expense of deodorants in the future. Often sharing these items of information of seemingly small significance to health personnel can help convey to the parents a sense that they are capable, cooperative partners in the provision of health care to their children and can diminish the feeling of distance between consumers and providers of health care.

Diabetes

The National Diabetes Advisory Board noted in May 1982:

> Diabetes is a major health problem for American Indians . . . Diabetes in this population is almost entirely of the non-insulin-dependent type. However, it is occurring at an unusually early age, frequently in the teens and early twenties, with a resultant earlier onset of complications and excessive mortality in the early adult and midadult years. Age specific death rates for diabetes are three to four times higher for American Indians aged 25–54 years compared with the general population.
>
> The complications of diabetes are appearing with distressing frequency: amputation rates are 2 to 3 times the general U.S. population and renal failure rates are 20 times as high. According to a recent study in the Ute Indian tribe, the prevalence of diabetes is four times the statewide rate, and the rate of diabetic nephropathy is 43 times the rate for the diabetic population in Utah generally. Thirty percent of the outpatient visits in some facilities are for diabetes-related problems [US Department of Health and Human Services, 1982, p 39].

Diabetes has made its appearance among Indians with the consumption of a diet high in refined carbohydrates and fats and low in traditional foods, which provided more protein and roughage. At the same time, life has become more sedentary. The Pima have long been recognized as having the highest prevalence of diabetes of the non-insulin-dependent type. Recent studies have shown that normal and prediabetic Pima have fasting and stimulated insulin levels two to three times greater than Whites (Aronoff et al, 1977; Pettitt, 1982). Obesity in the Pima is not a strong risk factor for death, though it is a strong risk factor of diabetes. Only 1 case of diabetes was noted among the Pima in 1908 (Knowler et al, 1978); 21 cases were reported among the Pima in 1940 (Joslin, 1940). Beginning in the 1950s the prevalence was identified as high, and major complications were seen (Cohen, 1954; Dorf et al, 1976; Parks, Waskow, 1961; Inglefinger et al, 1976). Other Native American and Eskimo groups have a similarly high incidence of diabetes. A review of admissions of 50 Indian and 50 non-Indian patients to a hospital in South Dakota revealed that 25 of the Indian admissions were diabetes-related (Dean, 1976).

Implications With as many as 30% of clinic visits occurring for diabetes, the magnitude of the problem is inescapable. Health care providers need to be active case finders. They can perform urine tests for glucose and acetone on individuals they know are under stress, such as recent postoperative patients, those admitted for reasons other than diabetes, or those at risk generally. Another useful endeavor is participation in community health screening clinics, particularly tribal-initiated health fairs and teaching projects that actively involve indigenous health workers. Diabetic clinics for pregnant women have been helpful support groups in some areas. Various community groups, women's clubs, weight-loss groups, and church organizations may welcome speakers on the topics of preventive health, diabetes recognition, and diabetes management. Clinic visit waiting times might be used for teaching sessions. As part of school curricula, programs could be developed to teach children the early recognition signs of diabetes and health-promoting diets that can be used with available foods. Because trading posts frequently lack fresh produce, and homes sometimes do not have storage facilities, dietary adjustments have to be made.

A unique opportunity exists among the Oglala to incorporate into health teaching an activity that could easily be pursued into adulthood. Billy Mills, an Oglala Sioux, won the 10,000-meter race in the 1964 Tokyo Olympics. He has returned to the reservation to speak to young people's groups and to participate in running events. He is highly regarded. Often reservation schools have state division championships in long-distance running. If this popular activity could be developed into a lifelong pattern, it would contribute to better health. The interest in long-distance running is reflected in an increasing number of races and the appearance of T-shirts and trophies noting events and identifying participants.

The reemergence of Indian clubs in schools has increased traditional dancing among children; this also is an activity many could continue into adulthood. Individuals dancing weekly at an Indian club and practicing for dance events achieve a considerable amount of exercise.

Alcoholism and Fetal Alcohol Syndrome

Numerous authors have noted the rates of cirrhosis and alcoholism among Native American populations (Reichenbach, 1967; Kunitz et al, 1971; Whittaker, 1982). In 1950, an extensive study among the Standing Rock Sioux of South Dakota showed that 1 in 3 persons over the age of 15 drank excessively. This was an incidence of alcoholism double that of the White comparison sample. Twenty years later the death rate for alcoholic cirrhosis was 27.3 per 100,000 among Indians, five times higher than the 6.1 per 100,000 in all other races. Age-specific rates of cirrhosis mortality for Indian women were 128 per 100,000 compared with 37 per 100,000 for Black women and 15 per 100,000 for White women. The rates for Indian men of the same age were even higher (Whittaker, 1982). The autopsy evidence of cirrhosis among Indians in the Phoenix area is three times that expected among non-Indians (Reichenbach, 1967). The Apache and Hopi death rates from cirrhosis are two to four times the national rate (Kunitz et al, 1971).

Social reasons have been implicated as factors in the development of alcoholism, but a biologic difference in alcohol metabolism is probably more to blame (Aase, 1981; Umbriet, Ostrow, 1980). Many Native Americans share with Asian Mongoloids a modification in the breakdown of alcohol (Ewing, Rouse, Pellizzari, 1974; Wolff, 1973). See chapter 4 for more details.

American Indians are also at risk for fetal alcohol syndrome. Present data are incomplete, but they implicate cultural influences, fertility, patterns of alcohol consumption and abuse, and perhaps dietary and metabolic differences (Aase, 1981; Umbriet, Ostrow, 1980). The signs and symptoms of the disease include prenatal and postnatal growth deficiencies and short palpebral fissures. Perhaps the most significant concomitant of the disease is mental retardation; fetal alcohol syndrome is the third leading cause of mental retardation in the state of Washington (Umbriet, Ostrow, 1980).

Implications Health professionals should be aware of the symptoms of alcoholism as well as the clinical manifestations of fetal alcohol syndrome. They should diligently counsel childbearing populations about the risks of fetal alcohol syndrome. Films from the March of Dimes on fetal alcohol syndrome and pamphlets developed by the Indian Health Service deal with both alcoholism and fetal alcohol syndrome in culturally specific ways. Early diagnosis and early patterning training of the infant with fetal alcohol syndrome have been shown to improve the long-term outcome. More research is needed on the causes of the high incidence of alcoholism—biologic, cultural, genetic—and on methods for treating alcoholics and their affected offspring.

Growth and Development

The literature on growth and development patterns among Native Americans is sketchy. In 1906 on the Pine Ridge reservation, a physician reported that the average length of the Indian newborn was 19.5 in and its average weight was 7.5 lb. The Indian baby was longer and heavier than the average infant born to White parents in the United States. "It appears the Indian infant begins life with a little better physique than does the white infant" (Walker, 1906). This same study indicated that the adult Indian male was 1.9 in taller than the average White male and weighed 6 lb more. Indians' average chest measurements were the same as White males' (Walker, 1906).

These measurement advantages have apparently been retained over time. A study that examined birth weight and body length found that three of the top four groups were Native Americans: in fourth place were Aleuts from the Aleutian Islands, with a mean weight of 7.4 lb; in third place the Sioux, with a mean birth weight of 7.5 lb and length of 19.6 in; and in first place the Cree, from James Bay, Canada, with a mean weight of 8.3 lb (Meredith, 1978). Two of the three groups at the top of the height and weight scale at age 9 were Native Americans: the Blackfeet (weight 63.5 lb, height 52.3 in) and the Cree (weight 62.3 lb, height 50.9 in) (Meredith, 1978).

Mongolian spots are common in American Indians and normally are present until age 3–4. The racial frequency of mongolian spots is discussed in the growth and development section of chapter 3. It is important to recognize their frequency in different populations, because health care providers can mistake them for trauma at birth or, more seriously, as signs of child abuse in small children.

Research about the growth and development of Native Americans is urgently needed. Norms should be established for specific groups. Greater motivation for both parents and health professionals might result from the use of growth charts unique to the population being evaluated.

Rh Blood Factor

While 15% of the non-Indian population is Rh negative, the percentage is much lower for American Indians, particularly as they approach full-blood quantum. Therefore, health personnel must be especially alert to obtain and administer Rh immunoglobulin to the Rh-negative mother who delivers an Rh-positive infant, to prevent the development of Rh sensitization and impairment of the health of future newborns. Rh immunoglobulin needs to be administered to the Rh-negative mother who undergoes any procedure or event that carries the potential of disrupting the placenta and permitting a fetomaternal exchange of blood cells. The most common possibility for occurrence of this is during amniocentesis. Every Rh-negative woman experiencing an abortion (spontaneous, therapeutic, or elective) or surgery for ectopic pregnancy must receive Rh immunoglobulin. Any event that increases the fetomaternal exchange of blood cells will require the administration of additional Rh immunoglobulin. Among these events are cesarean section, placenta previa, and abruptio placenta, as well as traumatic amniocentesis.

Arthritis

In 1969 attention was drawn to the increased frequency of chronic arthritis in Canadian Indian schoolchildren compared to the non-Indian population (Hill, Robinson, 1969; Rosenberg et al, 1982). In 1981 rheumatoid arthritis was found at a rate of 6.8% in a Chippewa band (Harvey et al, 1981). The highest prevalence recorded prior to this study was in the Yakima tribe in the state of Washington, whose women had a 3.4% incidence (Willkins et al, 1982). Prevalence of rheumatoid arthritis had been shown to range from 0.4% to 1.0% in the United States as a whole. A study among tribes in North America shows rates among the Blackfeet of 1.4%, Pima 1.8%, and Chippewa 6.8%, as mentioned above (Harvey et al, 1981). A genetic factor is implicated in the Chippewa study: 44% of the persons diagnosed belonged to two large kindreds (Harvey et al, 1981). The Navajo, who were previously thought to have a unique form of arthritis, have the same type as the general population but fail to demonstrate the usual blood reactive factors (Rate et al, 1980).

Implications The unusually high incidence of rheumatoid arthritis among segments of the Native American population and the findings in the Navajo of arthritis in the presence of negative blood studies should alert health care providers. Geographic location has been implicated as a factor, yet tribes that resided in one location and later moved do not demonstrate changed rates. Genetic factors have been implicated for the Chippewa, and this finding could direct further study. Health professionals alerted to this problem in Indians could make earlier diagnoses, thus preventing degenerative changes and the loss of confidence patients experience when genuine symptoms are brushed aside lightly.

Tuberculosis

Numerous investigators describe a high incidence of tuberculosis among Native Americans (Delien, 1951; Dublin, 1942; Heath, 1980; Reifel, 1949; Walker, 1906; Jones, 1940). The incidence for South Dakota today is 10 per 100,000, but the incidence on the Pine Ridge Reservation is 300 per 100,000 (Heath, 1980). World Health Organization (WHO) figures for 1974–1976 indicate that some areas of Africa, Asia, and Oceania have an annual incidence of 200–350 cases of tuberculosis per 100,000 (Mandell, 1975). These are the worst WHO figures. Thus, the Oglala Sioux have a problem with tuberculosis equal to the worst found in the Third World. Female morbidity from tuberculosis is higher than would be expected. This may be related to the high birth rate: lowered resistance during the postpartum period may contribute significantly to female tuberculosis morbidity rates (Heath, 1980).

Many of the studies relate the high rates among Indians to social and economic conditions. In 1912 rates among Indians were 1,040 per 100,000 while the overall U.S. rate was 145.5. By 1945 the rate was down to 211.9 among Indians and 40.1 in the U.S. population (Reifel, 1949). Rates vary among tribes from a low of 2.0% among the Apache to 4.6% among the Navajo. The rate for Pine Ridge, 3.2%, was exceeded only by Red Lake, Minnesota, Wind River, Wyoming, and the Navajo in Arizona (Reifel, 1949). In 1940, some 17% of the Oglala Sioux had chest lesions demonstrating far advanced stages of the disease (Alley, 1940).

Implications Because many authors implicate socioeconomic factors in tuberculosis, perhaps health personnel should participate more actively in matters not directly related to health. This can occur on a concrete basis by accepting assignments on tribal planning boards, especially when there is the opportunity to influence the design of facilities for large numbers of people, such as nursing homes and school residences. Because of the risk to the postpartum client in Pine Ridge, it is important that all health personnel who work with pregnant and postpartum clients be given thorough training about tuberculosis and its medical management. While in the hospital for delivery, all mothers should be given tuberculin skin tests, if not already known to be positive. These can be read prior to discharge. New individuals demonstrating Mantoux conversion should be evaluated prior to discharge, and medication should be begun if indicated. The state department of health currently provides a worker for tuberculosis follow-up. However, in Pine Ridge there is a long history of Native American distrust of the state. The tuberculosis control program would make a greater impact if the major outreach was conducted by currently employed tribal health workers—the community health representatives. They could provide culturally sensitive care without being distanced from the recipients by ethnic, racial, and social barriers.

Oral Clefts

The highest frequency of oral clefts in a racial group occurs among the Mongoloids (Niswander, Adams, 1967; Greene et al, 1964). The Japanese have the highest case rate, with 2.13 per 1,000 live births, followed by American Indians with 1.38 per 1,000 live births. Furthermore, the frequency of isolated cleft palate in American Indians is nearly identical to that of the Japanese and approximately twice that of the White and Black populations. Studies among a group of Montana Native Americans revealed a rate of cleft lip and palate of 1:276 (Tretsven, 1963). A genetic study of cleft lip and palate in Utah revealed support for the hypothesis that a genetic component exists for this anomaly, with a rate of 1:662 (Woolf, Woolf, Broadbent, 1963). The high frequency of oral clefts among the Flathead Indians in Montana suggests that a major gene has reached a high frequency in this population as a result of genetic drift (Niswander, Adams, 1967). More facial clefts occurred in boys, although isolated cleft palate was more frequent in girls (Greene et al, 1964).

Implications During periods of orientation to Indian Health Service hospitals, health personnel should receive information alerting them to the possibility of a greater frequency of oral clefts in Native Americans. In particular, staff in the labor and delivery, newborn, and postpartum areas should become knowledgeable in the care of children with oral clefts. They need to become skilled in dealing with the needs of parents faced with the crisis of having a child with a congenital anomaly. Personnel should be active in teaching community groups and ancillary health workers the biologic basis for the frequency of this anomaly in American Indian populations. Further research is needed to establish incidence rates for various Indian populations so that funds could be developed for genetic counseling, referral for corrective surgery, and establishment of specialty clinics on the reservations most affected.

Gallbladder Disease

Gallbladder disease has been widely noted among Native Americans (Brown, Christensen, 1967; Burch, Comess, Bennett, 1968; Kravetz, 1964; Comess, Bennett, Burch, 1967; Sievers, Marquis, 1962). In 1954 gallbladder disease was identified as the most frequently encountered surgical condition among American Indians operated upon at the Pine Ridge Hospital

(Lam, 1954). A 1972 report showed the Shoshone and Arapaho to have a much higher incidence than the U.S. population in general. The Indian population has a 2.2% incidence of gallbladder disease compared to a 0.3% incidence among the White population (Erickson, 1972). Previous studies have reported incidence rates of 3.6% in Sioux Indians, 2.3% in Pima Indians, and 1.2–1.5% in Navajos (Lam, 1954; Hesse, 1959; Gilbert, 1955; Salsbury, 1947). A more recent study indicates a substantial rise in gallbladder disease incidence among the Navajo/Hopi, from 1.2% in 1955 to 2.49% in 1972 (Gilbert, 1955; Porvaznik, 1972).

Comparative studies have been made on multiethnic groups in Hawaii, showing rates of gallbladder disease of 2.6% for Chinese, 2.1% for Puerto Ricans, and 1.8% for Portuguese (Yamase, McNamara, 1972). These studies indicate that the racial derivation of the Mongoloid peoples persists in a common pattern of disease incidence centuries later. See Table 8.1.

Implications The etiology of gallbladder disease remains uncertain despite numerous studies. It is known that childbearing and a high fat intake increase the incidence of the disease. For this reason, menus in schools, nursing homes, and the like should avoid high-fat diets while including items that satisfy cultural preferences. Health workers should be alert to symptoms of gallbladder disease and provide appropriate referral. Birth control pills should be used cautiously in Indian populations for many reasons but in this instance because they alter the composition of bile and affect its transport through the biliary tract. Education of health personnel about this disease is important. Because of the early age of onset in Indians, symptoms may be overlooked by uninformed practitioners. Community education about the significance of a positive family history, childbearing, and

TABLE 8.1 *Gallbladder disease incidence*

Racial group	Percentage incidence
Sioux	3.3
Chinese	2.57
Navajo/Hopi	2.49
Pima	2.3
Shoshone/Arapaho	2.2
Puerto Rican	2.08

Sources: Derived from Lam, 1954; Yamase, McNamara, 1972; Porvaznik, 1972; Hesse, 1959; Erickson, 1972.

a high-fat diet are again important. If the community is aware of signs and symptoms of gallbladder disease, patients are more likely to come for treatment early in the disease process and thus avoid the hazards of emergency surgery.

Phenylketonuria and Cystic Fibrosis

These two biologic variations are grouped together because at one time they were thought to be absent in Native American populations. Although rare, cystic fibrosis and PKU have been identified in Indian children (Wagner, Littman, 1967; Harris, Riley, 1968). The significance of these variations is that they must not be arbitrarily dismissed as potential diagnoses without appropriate screening and clinical investigation, simply because the individual is Native American.

Hyperbilirubinemia

The predisposition of American Indian and Eskimo populations to have higher bilirubin concentrations than Whites appears to be linked to their common hereditary relationship. Because of the higher bilirubin levels among Indian populations, it is important for clinicians to know that breast-feeding is associated with significantly higher total bilirubin levels only in White neonates (Postl, Nelson, Carson, 1982).

The higher bilirubin levels occurring among infants of Oriental and Native American descent also extend to Eskimo infants. Bilirubin levels peak in Eskimo neonates usually a day later than in Whites and Blacks (that is, on days 4 and 5) and remain elevated for a longer period. In Eskimo mothers this was associated with a high level of the ability of the breast milk to inhibit hepatic bilirubin glucuronyl transferase activity (Fisher et al, 1978). A study done specifically on the Sioux from Pine Ridge and Rosebud, South Dakota, indicated that infants that had one Indian parent and one White parent had bilirubin activity like that of Indian infants (Trygstad, McCracken, 1966).

Implications The protective effects of breast-feeding against bacterial infections are so remarkable that every effort should be made to support this method of infant feeding. A crucial factor in the continuation of breast-feeding is support by the hospital staff. It is poor practice to discontinue breast-feeding routinely when the Indian infant's bilirubin reaches 10 mg/100 ml. The mother who is asked to abstain from breast-feeding for even 24 hours infers that there is something harmful about her milk. Clinicians can often provide reassurance and ensure proper medical management by simply explaining the causes of high bilirubin levels in the newborn and the effectiveness of breast-feeding in helping the infant cope with these levels.

Breast-feeding is a *learned* art even among groups we assume to be more attuned to a natural way of life. Virtually no cultural group in this country has been untouched by the erroneous notion that bottle-feeding is a superior method of infant feeding. Health personnel now must overcome resistance to breast-feeding by reeducating the population.

Summary

The high incidence rates of gallbladder disease, lactose intolerance, rheumatoid arthritis, hereditary polymorphic light eruption, otitis media, and hyperbilirubinemia among American Indians warrant that health professionals familiarize themselves with these diseases. Health care implications of each have been discussed in this chapter. Knowledge of these diseases should alert researchers to the potential for other diseases and normal biologic variations in Native Americans.

The acknowledgment of biologic variation in no way diminishes our common humanity; it only makes it more fascinating. As Anshen comments, the unity of human beings must not be overlooked in the study of biologic or cultural diversity (Anshen, 1959). Rene Dubos points out the uniqueness of humans in this way: "Whereas other living things survive through adaptive changes in their bodies and their instincts, man strives to impose his own directional will on the relations that he has with the rest of the world" (Dubos, 1959, p 266).

It is when health care providers discover a people's adaptive changes that biologic variation becomes meaningful to the design of health care programs, to the delivery of health care, and to the promotion of a higher level of health than previously experienced.

REFERENCES

Aase J: The fetal alcohol syndrome in American Indians: A high risk group. *Neurobehav Toxicol Teratol* 1981; 3(2):153–156.

Alley R: Tuberculosis among Indians. *Dis Chest* 1940; 6(1):44–50.

Anshen R: Epilogue: World perspectives. Pages 283–292 in: *The Mirage of Health*. Dubos R (author). Harper & Row, 1959.

Aronoff S et al: Unexplained hyperinsulinemia in normal and "prediabetic" Pima Indians compared with normal Caucasians. *Diabetes* 1977; 26(9):827–840.

Bayless TM et al: Lactose and milk intolerance: Clinical implications. *N Engl J Med* 1975; 292(22):1156–1159.

Birt A: Photodermatitis in Indians of Manitoba. *Can Med Assoc J* 1968; 98(7):392–397.

Birt A, Davis RA: Hereditary polymorphic light eruptions of American Indians. *Int J Dermatol* 1975; 14:105–111.

Brandon W: *The American Heritage Book of Indians*. American Heritage, 1974.

Brody JA, Overfield T, McAlister R: Draining ears and deafness among Alaskan Eskimos. *Arch Otolaryngol* 1965; 81(1):29–33.

Brown JE, Christensen C: Biliary tract disease among the Navajos. *JAMA* 1967; 200:138–140.

Brues A: *People and Races*. Macmillan, 1977.

Burch T, Comess LJ, Bennett PH: The problem of gallbladder disease among the Pima. Pages 80–88 in: *Biomedical Challenges Presented by the American Indian*. Vol 165. Pan American Health Organization, 1968.

Caskey DA et al: Effects of age on lactose nonabsorption in Oklahoma Native Americans as determined by breath H_2 analysis. *Am J Dig Dis* 1977; 2:113–116.

Chaudhuri J, Chaudhuri J: Emerging American Indian politics: The problems of powerlessness. Pages 103–112 in: *Chicanos and Native Americans*. delaGarza R, Kruzewski A, Arciniega T (editors). Prentice-Hall, 1973.

Cohen EM: Diabetes mellitus among Indians of the American southwest: Its prevalence and clinical characteristics in a hospitalized population. *Ann Intern Med* 1954; 40:588–599.

Comess LJ, Bennett PH, Burch TA: Clinical gallbladder disease in Pima Indians. *N Engl J Med* 1967; 277(17):894–898.

Dean R: A comparison of health problems in comparable groups of Indian and White adults. *S Dak J Med* 1976; 29:7–13.

Delien H: Continuity of program—a necessity of tuberculosis control among American Indians. *Lancet* 1951; 71(4):136–137.

Dorf A et al: Retinopathy in Pima Indians. *Diabetes* 1976; 25(7):554–560.

Dublin L: The mortality from tuberculosis among the racial stock in the southwest. *Am Rev Tuberc* 1942; 45:61–74.

Dubois R: *The Mirage of Health*. Harper & Row, 1959.

Duncan IW, Scott EM: Lactose intolerance in Alaskan Indians and Eskimos. *Am J Clin Nutr* 1972; 25:867–868.

Erickson WG: Cholecystectomy and the American Indian. *Rocky Mt Med J* 1972; 69:33–35.

Everett M et al: Light sensitive eruptions in American Indians. *Arch Dermatol* 1961; 83:243–248.

Ewing JA, Rouse BA, Pellizzari ED: Alcohol sensitivity and ethnic background. *Am J Psychiatry* 1974; 131(2):206–210.

Fisher Q et al: Jaundice and breast feeding among Alaskan Eskimo newborns. *Am J Dis Child* 1978; 132:859–861.

Fusaro R, Johnson J: Hereditary polymorphic light eruption in American Indians. *JAMA* 1980; 244(13):1456–1459.

Gilbert J: Absence of coronary thrombosis in Navajo Indians. *Calif Med* 1955; 82:114.

Greene J et al: Epidemiologic study of cleft lip and cleft palate in four states. *J Am Dent Assoc* 1964; 68:387–404.

Gregg JB, Steele JP: Mastoid development in ancient and modern populations. *JAMA* 1982; 248(4):459–461.

Griffin J: The origin and dispersion of American Indians in North America. Pages 11–16 in: *Biomedical Challenges Presented by the American Indian.* Vol 165. Pan American Health Organization, 1968.

Harris R, Riley H: Cystic fibrosis in the American Indian. *Pediatrics* 1968; 41:733–738.

Harvey J et al: Rheumatoid arthritis in a Chippewa band. *Arthritis Rheum* 1981; 24(5):717–721.

Hassrick R: *The Sioux*. University of Oklahoma Press, 1964.

Heath C: *A Descriptive and Evaluative Study of the Tuberculosis Occurring in American Indians Residing in Shannon and Washabaugh Counties, South Dakota, 1970 through 1978*. (Thesis.) University of Texas, Houston, 1980.

Hesse FG: Incidence of cholecystitis and other diseases among Pima Indians of southern Arizona. *JAMA* 1959; 170(15):1789–1790.

Hill RH, Robinson HS: Rheumatoid arthritis and ankylosing spondylitis in British Columbia Indians. *Can Med Assoc J* 1969; 100:509–511.

Hongladarom GC, Russell M: An ethnic difference—lactose intolerance. *Nurs Outlook* 1976; 24(2):764–765.

Hyde G: *Red Cloud's Folk*. University of Oklahoma Press, 1976.

Inglefinger JA et al: Coronary heart disease in the Pima Indians. *Diabetes* 1976; 25(7):561–565.

Johnson RL: Chronic otitis media in school age Navajo Indians. *Laryngoscope* 1967; 77:1990–1995.

Jones LR: Tuberculosis on a small thickly populated Indian reservation. *Am Rev Tuberc* 1940; 42:197–202.

Jorgensen JG: A century of political economic effect on American Indian society. *J Ethnic Studies* 1978; 6:1–83.

Josephy AM: *The Indian Heritage of America*. Bantam Books, 1968.

Joslin EP: The universality of diabetes. *JAMA* 1940; 115:2033–2038.

Knowler WC et al: Diabetes incidence and prevalence in Pima Indians: A 19-fold greater incidence than in Rochester, Minnesota. *Am J Epidemiol* 1978; 108:497–505.

Kravetz R: Etiology of biliary tract disease in southwestern American Indians. *Gastroenterology* 1964; 46(4):392–398.

Kunitz SJ et al: The epidemiology of alcoholic cirrhosis in 2 southwestern Indian tribes. *Q J Studies Alcohol* 1971; 32:706–720.

Lam R: Gallbladder diseases among the American Indians. *Lancet* 1954; 74:305–309.

Lebenthal E: Lactose malabsorption and milk consumption in infants and children. *Am J Dis Child* 1979; 133:21–23.

Mandell R: *Principles and Practices of Infectious Diseases*. Appleton-Century-Crofts, 1975.

Matsunaga E: The dimorphism in human normal cerumen. *Ann Hum Genet* 1962; 25:273–286.

Maynard E, Twiss G: *That These People May Live*. DHEW Publication No. HSM 72–508. US Department of Health, Education, and Welfare, 1970.

Meredith H: *Human Body Growth in the First Ten Years of Life*. State Printing Co., 1978.

Newcomber AD et al: Family studies of lactose deficiency in the American Indian. *Gastroenterology* 1977; 73(5):985–988.

Niswander J, Adams MS: Oral clefts in American Indians. *Public Health Rep* 1967; 82(9):807–812.

Parks JH, Waskow E: Diabetes among the Pima Indians of Arizona. *Ariz Med* 1961; 18:99–106.

Petrakis N, Molohon K, Tepper D: Cerumen in American Indians: Genetic implications of sticky and dry types. *Science* 1967; 158:1192–1193.

Pettitt D: Mortality as a function of obesity and diabetes mellitus. *Am J Epidemiol* 1982; 115(3):359–366.

Porvaznik J: Surgical problems of the Navajo and Hopi Indians. *Am J Surg* 1972; 123:545–548.

Postl BD, Nelson N, Carson J: Hyperbilirubinemia in Inuit neonates. *Can Med Assoc J* 1982; 126:811–813.

Rate R et al: Navajo arthritis reconsidered. *Arthritis Rheum* 1980; 23(11):1299–1302.

Reed D, Struve S, Maynard J: Otitis media and hearing deficiency among Eskimo children: A cohort study. *Am J Public Health* 1967; 57(9):1657–1665.

Reichenbach DD: Autopsy incidence of diseases among southwestern American Indians. *Arch Pathol* 1967; 84:81–86.

Reifel A: Tuberculosis among Indians of the United States. *Dis Chest* 1949; 16:234–249.

Rosenberg A et al: Rheumatic diseases in western Canadian Indian children. *J Rheumatol* 1982; 9(4):589–592.

Salsbury CG: Incidence of certain diseases among Navajos. *Ariz Med* 1947; 4(6):29–31.

Schenck R: Controlled trial of methoxsalen in solar dermatitis of Chippewa Indians. *JAMA* 1960; 172(11):1134–1137.

Sievers ML, Marquis M: Southwestern American Indians' burden, biliary disease. *JAMA* 1962; 182(5):570–572.

Spencer R, Jennings J: *The Native Americans*. Harper & Row, 1965.

Stuart P: The reservation community. *Pine Ridge Res Bull* 1969; 7:67–82.

Timmermans FJ, Gerson S: Chronic granulomatous otitis media in bottle-fed Inuit children. *Can Med Assoc J* 1980; 122:545–547.

Tretsven VE: Incidence of cleft lip and palate in Montana Indians. *J Speech Hear Disord* 1963; 28:52–57.

Trygstad CW, McCracken CH: The pattern of physiologic jaundice in the Sioux. (Abstract.) *Soc Pediatr Res* 1966; 5:86.

Umbriet J, Ostrow L: The fetal alcohol syndrome. *Ment Retard* 1980; 18(3):109–111.

US Department of Health and Human Services, Public Health Service, National Institutes of Health: *Diabetes in the 80's*. Report of the National Diabetes Advisory Board. US Government Printing Office, 1982.

Wagner M, Littman B: Phenylketonuria in the American Indian. *Pediatrics* 1967; 39(1):108–110.

Walker JR: Tuberculosis among the Oglala Sioux. *Am J Med Sci* 1906; 122:600–605.

Whittaker J: Alcohol and the Standing Rock Sioux tribe. *J Stud Alcohol* 1982; 43(3):191–200.

Willkins R et al: HLA antigens in Yakima Indians with rheumatoid arthritis. *Arthritis Rheum* 1982; 25(12):1435–1439.

Wissler C: *Indians of the United States*. Doubleday, 1940.

Wolff PH: Vasomotor sensitivity to alcohol in diverse Mongoloid populations. *Am J Hum Genet* 1973; 25:193–199.

Woolf CM, Woolf RM, Broadbent TR: A genetic study of cleft lip and palate in Utah. *Am J Hum Genet* 1963; 15:209–215.

Yamase H, McNamara JJ: Geographic differences in the incidence of gallbladder disease. *Am J Surg* 1972; 123:667–670.

Afterword

This is a small book for good reason: biologic variation is relatively new to health care workers; consequently, they have not studied it in depth or in a systematic way. The present book indicates the paucity of data, both by its size and by the lack of replication and concentration in many areas. In some cases only one study was found that addressed a topic. Reliance on a single study is not good, but it is better to cite one well-done study than to ignore the topic. Poorly executed or purely anecdotal studies have not been mentioned; only those judged to be scientifically worthwhile are cited.

In many cases the cited material represents the total available literature on the subject. Obviously much more research is needed. A good deal of the available data were gathered many years ago; it is probably time to redo many of these old studies.

The sparseness of data on the subject, in general, is reflected in the paucity of information on specific groups. Hispanic Americans, many Native American tribes, and Oriental Americans (to name some major groups) have not been studied much at all. The best available information is on Black-White comparisons and that is by no means complete. Specific areas of biologic variation also need much more research—for instance, growth and development, aging, dietary effects, clinical measurements, biochemical variants, and nervous system differences.

Those who work with specific racial groups need to search the scientific literature for information that will help them provide racially sensitive health care, and they should share this information with other health workers. The chapter on Sioux Indians is an example of this approach.

It is time for biologic variation to become an important subject for health research. Clinicians and their counterparts in academe need to collaborate on studies that elucidate the yet undiscovered variety in people, especially the variety that affects the delivery of health care.

Index